A Place
on the
Glacial Till

A Place

on the

Glacial Till

*Time, Land, and Nature
within an American Town*

Thomas Fairchild Sherman

Drawings by Byron Fouts

New York   Oxford
OXFORD UNIVERSITY PRESS
1997

Oxford University Press

Oxford   New York
Athens   Auckland   Bangkok   Bogotá   Bombay
Buenos Aires   Calcutta   Cape Town   Dar es Salaam
Delhi   Florence   Hong Kong   Istanbul   Karachi
Kuala Lumpur   Madras   Madrid   Melbourne
Mexico City   Nairobi   Paris   Singapore
Taipei   Tokyo   Toronto

and associated companies in
Berlin   Ibadan

Copyright © 1997 by Oxford University Press, Inc.

Drawings by Byron Fouts

Published by Oxford University Press, Inc.,
198 Madison Avenue, New York, New York 10016

Oxford is a registered trademark of Oxford University Press

Library of Congress Cataloging-in-Publication Data
Sherman, Thomas Fairchild.
A place on the glacial till : time, land, and nature within an
American town / Thomas Fairchild Sherman ; drawings by Byron Fouts.
p. cm.   Includes bibliographical references.
ISBN 0-19-510442-0
1. Geology—Ohio—Oberlin Region. 2. Geomorphology—Ohio—
Oberlin Region. 3. Ecology—Ohio—Oberlin Region. I. Title.
QE152.O24S54   1996
917.71'23—dc20   96-806

9 8 7 6 5 4 3 2 1

Printed in the United States of America
on acid-free paper

For Kátia...

Bright warbler from a sunlit world
A gift of springtime to unite
With truth and music of distant shores
The weathered continents of my life.

Contents

Drawings

Maps

A Place

on the

Glacial Till

There is a river called the Vermilion near Oberlin, Ohio. Its waters flow into Lake Erie and, after a year or two in that expansive but shallow basin, plunge over Niagara Falls, to Lake Ontario, the St. Lawrence, and the Atlantic. About 4 miles south of Lake Erie, a tiny brook called Chance Creek finds its way to the river, emerging from a narrow gorge overhung with hemlock and hay-scented fern to be greeted, across the river, by a broad floodplain decked with cottonwood and sycamore. Where the waters meet, one can sit on the shale ledges of a narrow promontory and contemplate the conjunction of worlds and the flow of time.

In our hours of thoughtless youth, time hardly flows with the river. I played as a child on the shale ledges above the falls of Ithaca, and wondered if a great slab of rock, split loose at the top, would ever fall. I knew nothing of ice ages and erosion, or of mastodons that once roamed the high shores of a different world, a world whose stones I then held in my hand. The tall European poplars that lined my American street seemed always to have stood as giants there. I did not think of the world as very old or, for all its age, as very new. There was fascination in the fast tumbling water, but the slow cutting of soft rocks went everywhere unseen.

The earth itself has a life like ours, though it grows young while growing old. Each day dawns with the face of a child, for dawn itself is ever present at the turning interface of time. The breezes stirring in the white pine above, the waters flowing below, the rising mists and falling rain, and all the motions of a million living things derive from a steady, central sun; yet always the energizing rays fall on a changing, creative scene.

Our ideas on these things too are young and ever-changing. Not for long have we viewed the earth as shaped by the same humble forces that work in our everyday lives. Three hundred years ago, what daring spirits could agree with Nicolaus Stensen (in Denmark) or Robert Hooke (in England) that all the layered rocks of the earth—from the shale in the gorges of New York and Ohio, to the chalk downs of England and the dolomites of the Alps—were formed once as sediments falling out of water, as dust settles on the windowsills of our home? Stensen said that lower layers must be older than those above, as with the black Devonian shale beneath the red Mississippian shale of the Vermilion gorge. The black rocks before me in this Ohio gorge carry the name of the county of Devon in England, for the sun that rose 340 million years ago saw sediments accumulating both here and there. That rocks of similar age could be found in different parts of the world has been known only since the early nineteenth century, when William Smith, helping canal builders find their way through the rocks of southern England, noticed that strata could be identified by their fossils. Snails and fish of hundreds of millions of years ago have kept a record to be read in the library of nature. All places of the world

4

have shared a journey that has made the earth our home; to listen from any spot is to hear the quiet echoes of a billion cycles around the sun.

Some of the most spectacular fish fossils in the world have come from the Devonian shale of the Black River and the Rocky River, the next two major streams east of the Vermilion. They are fossils of huge, heavily armored placoderm fishes that were swimming in this inland sea now beneath our feet. Their armor may have protected them from freshwaters as well as from fierce opponents, allowing their salty bodies to explore, without osmotic flooding, the rivers flowing into the sea. Now their remains lie in shale beneath our cornfields, woods, and homes, surfacing only where today's rivers and lake have cut into the sediments of their ancient world.

The streams and lake, which seem so old, are 10,000 times younger than the rocks that edge their waters. The Vermilion River did not begin to flow until the ice of the continental glacier retreated from northern Ohio, and in just 12,000 years the river has cut 100 feet through strata that had survived intact for 300 million years. The river's work did not go unseen. For more than 2,000 years, people have planted fields and lived beside it, and for many millennia it has been used, with the other streams of northern Ohio, for traveling between Lake Erie and the flourishing villages to the south along the Ohio River and its tributaries. One mile upstream from Chance Creek is a site high above the Vermilion River where Native Americans farmed and lived 2,500 years ago, in the time of Socrates and Plato.

The American land has a human history that goes back not just to the time of European settlements, but at least 11,000 years before that. Agriculture is as old in the New World as it is in England. The land, rivers, and lakes had names long before they were "discovered" by people speaking Spanish and French,

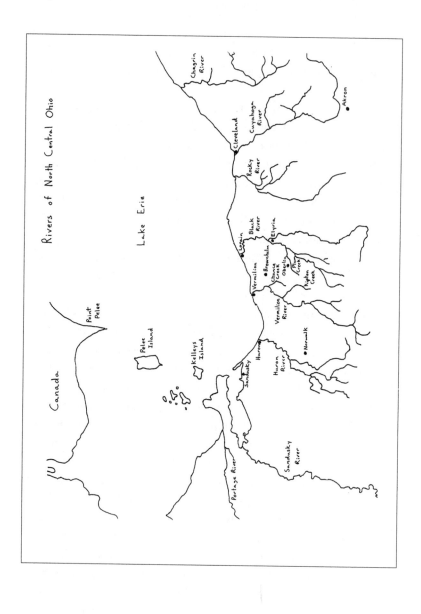

Rivers of North Central Ohio

Dutch and English; forests had been managed by stone axes and fire, and fields by the hoe. For thousands of years, the land has been shaped by human as well as natural forces. The leaves that float down our streams today are those of trees and herbs that have reached our land at various times after the glacial ice melted; some were slowly brought north as seeds by wind or birds or mammals, some came with the commerce and migrations of the native people, and many crossed the Atlantic in recent centuries from the fields of Europe and Asia.

In the years following the American Revolution, this land was known in Virginia as the Ohio Country, and in New England as the Connecticut Western Reserve. When settlers first crossed the Appalachians into southern Ohio from Virginia, or into the North from New England, the land was a heavily forested wilderness, because few people still lived in this area. Since the early seventeenth century, and perhaps even before, the Native Americans had been ravaged by European diseases and by the ecological, social, and political upheavals wrought by contact with people from Spain, France, Holland, and England. The first New England farmers came to the forests of Lorain County in 1807. In less than fifty years, nearly half the forests had been cut, and by 1940 only 7 percent of the county was still covered by woodlands.

When I came to this land as a student in the 1950s, I was employed in the summers as a wilderness guide in the Quetico–Superior country of Canada and Minnesota. The woods and waters of that northern country, seen from a canoe slipping between the clouds of the sky and those reflected in the water, past shores of granite ledges and pines, created for me a new image and ideal of beauty. Here were scenes even more beautiful than the waterfalls of my home, for people had not diverted any of the waters or distracted nature's ways. Wilderness became the standard against which all natural beauty could be measured; a land was beautiful to the extent that it approached its natural condition. I might have thought that the county of my college was about 7 percent as beautiful as it would have been if farmers had not cut down its trees. Then I went to England for four years to live and study among the medieval buildings and gardens of Oxford. In Britain there is no wilderness, no place that has not been influenced by thousands of years of human activity—but there is exquisite beauty. In the Cotswold hills north of Oxford is the little village of Churchill, to which I would retreat when my college closed for vacation. It was the birthplace of the geologist William Smith, who from that tiny hamlet showed how the history of our whole earth could be better understood. From lanes winding out of the village across the countryside one could see horses pasturing in fields still shadowed by the rolling ridges and furrows created by medieval plows. A day, in all its green and pleasant ways, was all its yesterdays as well. Perhaps the beauty of nature can take on an even deeper meaning where wilderness is embraced by human love and aspirations and where the land records a poignant hymn of the "still, sad music of humanity."

I have walked through beech woods in England similar to the beech and maple woods that are native to much of Ohio. A favorite English beech wood hangs from a hill above the tiny village of Selborne in Hampshire. Yew trees (from which the English long bow was once made) join the beech on the Selborne hillside, looking much like the hemlocks among the Ohio beech. Many flowers are closely similar to ours, and several of the ferns are identical. The thatched roofs on the cottages are made from a reed, *Phragmites communis*, that grows in our wetlands too. In July 1776, a few days after the signing of the Declaration of Independence in Philadelphia, General John Burgoyne's cavalry rode through this village toward Portsmouth and a war in America. But history remembers Selborne for the man who saw the troops pass: Gilbert White, who delighted in watching everything that touched the life of his native village and its surrounding countryside, but who noted especially the ways of its birds and other animals and of its plants. The village today is much as it was 200 years ago, and has a sense of the sanctuary about it. One cannot walk its lanes without believing that it represents a harmony between people and the land, where humanity and nature are so intimately mixed that village and woods and fields are all one. The wilderness that once was there has yielded to a new and intimate beauty, a beauty not only of the land but of the people who cherish it. The wilderness is within.

There is a wilderness of time and nature within any place on earth. When in the nineteenth century many New Englanders made the long trip to northern Ohio, Henry Thoreau walked the few miles to Walden Pond. It was one of Thoreau's great rules of life that any Walden Pond contains the reflections of the entire world within it—that one can see and know and feel, if eye and mind and heart be open, more at one's own doorstep than hurried travels will ever reveal in the far corners of the earth. Thoreau longed to know "the entire poem" of heaven and earth revealed in the reflections at Walden, as Tennyson yearned to know his flower in the crannied wall. Words and verses of the poem are everywhere before us—at Selborne, or at Walden, or in the woods and fields and villages of Ohio.

1 rocks

The land of the southern Great Lakes lies as a soft carpet on the rocky core of North America. As I look out on the lush green lawns of May in my village, rich with the spectral colors that their tiny flowers—veronica and violets, dandelions and English daisies—disperse from the radiance of the sun, I hear from the leafing boughs above the busy songs of the warblers and woodpeckers, the robins and cardinals, and feel the earth speaking of a day that has never been before. This land, with its myriad threads of living things, recently woven from deeper time and many distant places, wraps us in a world of song and freshness above the silence of aeons we have never known.

Rocks are the symbol of earth's enduring structure, yet it is the rocks that tell us of eternal change in the development of our land and of those who have dwelt on it. Below the green carpet of northern Ohio lie thousands of feet, and hundreds of millions of years, of accumulated sediments containing the rocky impressions of the living music and colors that greeted

11

the May mornings of earlier orbits around the sun. Beneath those hardened clays and sands and lime lies the sunken core of North America, the granites and other igneous rocks that reach the surface of our continent in the great Canadian Shield to our north.

In the rocks of the earth we can hold in our hands the crystals of nature's creative spirit, an elemental architecture of energies that have shaped our world since the beginning of time. Awe is inspired by the mysterious hues and patterns of an agate or an amethyst. For me the Canadian granites, like the most ancient rocks far beneath my home, provoke a special sense of wonder. During youthful summers near Algonquin Park in Canada, I would struggle with my canoe 2 miles up a primitive portage trail to reach what seemed the highest and most remote water beneath the sky. We called it Rock Lake, and on its stony ledges we would lie awake far into the night, as sleep came slowly to our tired bodies. The night suspended us in a galactic darkness, a silence between water and stars, when from a corner of eternity, as from the aisles of an ancient cathedral, the eerie cry of a loon proclaimed the divinity of the Milky Way. Like that great northern diver, at one with the Atlantic and the boreal lakes, we can touch the rocks and waters of the earth and sense the powers that have bestowed this home within the stars.

The rocks of northern Ohio are mostly out of sight, covered by many feet of soil. For several miles southward from Lake Erie, the soils are "lake till," deposited as sediments when the lake was higher than it is today. South of the sandy beach ridges delimiting the highest reaches of the postglacial lakes are the clay soils of the "glacial till," dropped from the melting ice of the continental glaciers that once rode heavily on this land. On the shores of Lake Erie, in the gorges of our streams as they approach the lake, and sometimes in the upper reaches of those streams where the till is thin, the rocks lie exposed. Elsewhere one can dig in the garden all day without a thought to the rocky foundations that are holding up our world. How different this is from the mountains of Maine and Colorado or the rocky shores of the Canadian lakes. Our trees can root deeply in our woodlands and gardens.

For hundreds of feet beneath the glacial till, the rocks of our area are almost entirely shale and sandstone, though limestone and dolomite surface a little to our west. An open sandstone quarry at South Amherst, 160 feet deep, is the largest in the world. All these rocks are sedimentary: they had their origin as sediments accumulating on the bottom of a body of water. Their presence means that this land, now 800 feet above sea level, was once covered by water, probably by the ocean itself. The great thickness of these sedimentary rocks means that this "land" was a sea for a very long time.

Here and there in our fields and woods are isolated rocks of a different kind, lying in or on the clay soils of the glacial till. They are granites and other hard igneous rocks forged by the heat of the earth's interior, Precambrian rocks more ancient than any of the sedimentary rocks beneath the till. They come in all sizes, and in the streambeds of the Vermilion River and Chance Creek, I sometimes find a great clump of them that has fallen into the eroded gorge from a glacial moraine that was once above. I celebrate the presence of each one and remember the nights on Rock Lake. For these are pieces of the Canadian Shield, plucked off by the glaciers and carried south to our land. These "glacial erratics" deserve an awe and wonder for the forces that have brought them to the surface of our land from deeper and distant origins a billion years ago. The early Americans bestowed on them a spiritual meaning and sometimes (as at Red Rocks, Minnesota) painted them with the red ocher pig-

Geologic Eras and Periods

Era	Period	Millions of Years Before Present
Cenozoic	Quaternary	2.5–present
	Tertiary	65–2.5
Mesozoic	Cretaceous	135–65
	Jurassic	190–135
	Triassic	225–190
Paleozoic	Permian	280–225
	Pennsylvanian	320–280
	Mississippian	345–320
	Devonian	400–345
	Silurian	440–400
	Ordovician	500–440
	Cambrian	570–500
Precambrian		4,600–570

ments that symbolized enduring life. In Oberlin, three large erratics have been moved from their glacial resting places to stand as commemorative stones on Tappan Square.

Deep wells drilled into the bedrock of northern Ohio have shown that about 5,000 feet of sedimentary rocks lie beneath the glacial till, overriding a foundation of igneous Precambrian rocks similar to those of the Canadian shield or to those that rest on Oberlin's village square. This "geologic column" of sedimentary strata contains, from the top down, rocks from the Mississippian (Carboniferous), Devonian, Silurian, Ordovician, and Cambrian periods.

Our local rocks have global names. The names designate periods in the lifetime of our planet, and hence describe the relative age of the rocks, but not their location, form, or composition. The names were assigned, not without controversy, by

English geologists exploring Britain, where by extraordinary good fortune nearly every major period in the earth's history is represented by rocks exposed somewhere on the surface. As it happens, the oldest rocks were named first, during the 1830s, while Oberlin was taking shape as a college campus. If American geologists had been on the scene just a little earlier, our "Devonian" rocks might be called "Cayugan" or perhaps "Cuyahogan," for Devonian rocks are much more widespread in New York, Pennsylvania, and Ohio than they are in Devonshire or anywhere else in England. The rocks immediately above the Devonian were named the Lower and Upper Carboniferous, for they are in Britain the most important coal-bearing strata, but American geologists prefer to call them the Mississippian and Pennsylvanian rocks.

In 1831 (two years before Oberlin was settled), a young theology student, Charles Darwin, joined Adam Sedgwick on a fossil-hunting expedition in North Wales. Sedgwick, who was professor of geology at Cambridge University and president of the Geological Society, had been greatly influenced by William Smith's recent discovery that the rocks of England could be classified and arranged in age by the fossils they contained. As Sedgwick and Darwin hiked into the mountain cirque (or "corrie" or, in Welsh, "cwm") at Idwal in Wales, they could have had little inkling that they were going to change, in the immediate years ahead, our views of all land and nature and time. Darwin was about to embark on the *Beagle* on a worldwide voyage that would open his mind, in South America and the Galápagos Islands, to a nature that changed continually through time and space.

Sedgwick was beginning to portray the aeons of time that had formed the rocks of Britain and the world. In Wales, he was studying what appeared to be the oldest (lowermost) fossil-bearing rocks in Britain, and in 1835 he named them the "Cambrian." Cambria was the Roman name for Wales (derived from the Welsh "Cymru"), but no doubt it also had an attractive ring to a Cambridge man. So the Romans, the Welsh, and the professor from Cambridge gave the name to those Cambrian rocks that lie a mile deep under our Ohio village.

In Sedgwick's and Darwin's time, no one could know how old the Cambrian rocks are, but modern studies of minute amounts of naturally occurring radioactive isotopes suggest that they are 500 to 570 million years old. Cambrian rocks contain fossil representatives of all the present-day phyla of invertebrate animals, as well as fossils of algal plants. All the Cambrian animals and plants were probably confined to the oceans, with no life yet inhabiting the freshwaters and dry land of the earth. In the Burgess Shales of British Columbia are the Cambrian fossils of many animals that do not fit into any of our modern groups, suggesting that the oceans of 500 million years ago were more diverse in their animal life than are the seas of our world today. Especially prominent were trilobites of many forms and sizes, making up about 70 percent of Cambrian fossils; all trilobites are now extinct.

The Cambrian period marks the beginning of diverse animal life on the earth, the start of the Paleozoic era, or Age of Invertebrates. Below the Cambrian is the Precambrian, extending back 3.5 billion years toward the birth of the earth. In Sedgwick's time, no fossils had been seen below the Cambrian, but today we have many Precambrian fossils of "procaryotic" bacteria and blue-green algae (small cells devoid of nuclei). The Precambrian origins of life on earth lie 2 or 3 billion years in the past, but it was not until Cambrian times that life exploded into a great diversity of multicellular organisms, made up of "eucaryotic" cells containing nuclei and capable of leaving fossils that could be easily seen with the naked eye. The Precambrian rocks below northern Ohio are probably about 1 billion years old, as are those of the southeastern part of the Canadian Shield; in the upper Great Lakes and farther northwest, much of the Shield is 2.5 billion years old. While many of the rocks of the Canadian Shield have been formed from molten material, or have been highly metamorphosed by high temperatures or pressures, yielding granites and granite–gneisses, Precambrian rocks can be (as at the Grand Canyon) sedimentary sandstones or limestones that have survived for 600 million years or more.

Above the Cambrian rocks are those of the Ordovician and Silurian periods. The Silurian was recognized in Britain a little before the older Ordovician. While Darwin and Sedgwick were exploring North Wales in 1831, Sedgwick's friend Roderick Murchison was in South Wales, studying a formation of graywacke (sandstone containing an abundance of clay) that extended through the upper part of Wordsworth's beloved Wye Valley. In 1835 Murchison named the period he was studying "Silurian" in honor of an ancient Celtic tribe that had lived in that region.

The Silurian period saw some animals and plants living in the freshwaters and perhaps even on land, though the greatest diversity of life remained in the oceans. North American Silurian rocks are mostly dolomites and limestones containing an abundance of coral and brachiopod fossils. They are an important part of all the Great Lakes basins south of Lake Superior, because they form a circle around the western and northern sides of Lake Michigan and Lake Huron, meet Lake Erie west of Sandusky and at the Niagara River, and extend along the southern shore of Lake Ontario. The walls of the Niagara gorge are one of the greatest displays of Silurian rocks in the world. A relatively hard Silurian limestone forms the cap of Niagara Falls, and hence the eastern rim of Lake Erie. When the falls cut away this Silurian rim (perhaps in 25,000 years), our shallow lake will become a river.

Great blocks of Silurian limestone have been dropped on the shore of Lake Erie north of Oberlin, to protect the railroad from the storms of the lake—"railroad erratics" we might call them. They contain an abundance of fossils, including large corals. During late Silurian times, an isolated body of salt water apparently filled a basin occupying much of lower Michigan, northeastern Ohio, western Pennsylvania, and the Finger Lakes of New York; salt precipitated in that basin is now mined beneath Lake Erie at Cleveland (and beneath Lake Cayuga in New York).

Between the Cambrian and Silurian strata, Ohio contains a large amount of Ordovician rocks, buried deep in our region, but surfacing at Cincinnati and in the southwestern part of the state. In Wales, these strata are less distinctive, and Sedgwick and Murchison could not agree on how to designate them; Sedgwick claimed them as part of the Cambrian, and Murchison as part of the Silurian. It was not until 1879 that the Ordovician was granted geologic independence and named for the Celtic people who once lived in North Wales.

In the Ordovician period, life was still confined to the earth's seas, but a few vertebrate fishes (jawless ostracoderms) were invading the Age of Invertebrates. Ordovician limestones contain an abundance of trilobites and brachiopods, and Ordovician shales display a large number of graptolites (traceries of tiny colonial animals).

Near the surface of bedrock in the Oberlin region are the Devonian strata; these shales and siltstones are exposed in the gorges of the Vermilion, Black, and Rocky rivers, and along the shore of Lake Erie. The Devonian was named by Sedgwick and Murchison from studies of the north and south shores of Devonshire in England, but the English geologists would have found it easier to characterize this period had they been able to study the Devonian rocks of America.

In Britain, Devonian rocks display a disarming diversity in mineral type and color, which initially suggested that they were not of a common period. In South Wales and Herefordshire, a colorful formation called the Old Red Sandstone lies in undisturbed layers containing fossils of creatures from freshwater and terrestrial environments. In Devonshire, just across the Bristol Channel from Wales, are dull graywacke rocks that look nothing like the Old Red Sandstone; they have been greatly disturbed by folding and contain fossils of marine organisms. Yet a careful study of both formations revealed that they had something in common: In each, the fossils of the lower layers resembled Silurian fossils, while those of the upper layers were similar to fossils found in Carboniferous rocks. Murchison and Sedgwick decided that these heterogeneous rocks represented a common Devonian period, though they had been formed from sediments settling in very different environments (freshwater in Wales and oceanic in England) and had experienced different geologic histories since their formation.

The Devonian rocks of Ohio do not exhibit such striking heterogeneities. They contain, however, some of the most extraordinary Paleozoic fish fossils in the world, some of which are on display in the Cleveland Museum of Natural History. The first fossils were found in natural exposures where the rivers and lake cut into the Devonian shale. Many more were recovered by David Dunkle of the museum during construction of Interstate 71 southwest of Cleveland—reminding us that these remarkable fish once swam everywhere beneath our feet. When fish fossils are found in river gorges and on lake shores, they do not seem out of place, but fish beneath our farms and highways demand a different view of the history of the land—as do sharks' teeth and seashells found high in the Alps, or dinosaur remains uncovered in Antarctica.

The prize of the Cleveland Devonian fish fossils is the huge placoderm *Dunkleosteos*, named for David Dunkle. The placoderm fishes were covered by armored plates, which may have been turtle-like protection from predators. But during the Devonian period life invaded the freshwaters and the land, giving rise to insects, amphibia, psilophytes, and ferns. The armor of the placoderms may have been protection against the osmotic difficulties presented by life in brackish water or freshwaters, where water would tend to penetrate the salty tissues of the fish in a way that it would not do in marine environments. Placoderm fish fossils were also found by Hugh Miller in the Old Red Sandstone of Britain. Because of its abundance of fish fossils, the Devonian is often known as the Age of Fishes, but it is also the time when vascular plants and vertebrate animals became established on dry land—though northern Ohio, with its placoderm fish, was still beneath the waters.

As Murchison and Sedgwick learned from the Devonian rocks of Britain, a rock's appearance does not indicate its age, nor does the age of a rock tell us how it will look. All ages can produce sedimentary rocks that are sandstones from accumulated sand, shales from accumulated clay, limestones or chalk from accumulated shells, or hybrids of these types. (All ages can also produce volcanic igneous rocks of various kinds, but none of these intrude into our area.) Sandstones or shales require that sand or clay be brought into a basin by flowing air or, more likely, water. A stream carrying both sand and mud will dump the larger sand particles before it dumps the clay, so that whether sand or clay is accumulated in a given place will depend on its position relative to the sources of these materials. Limestone or chalk is produced by living organisms (such as corals or molluscs) that bring carbonates together as part of their life structure. The accumulation of all sediments is likely to be most rapid in offshore shallows. The ancient waters that covered so much of the central part of North America from Cambrian to Carboniferous times were probably very shallow. The thickness of the deposits, which extend far below the present-day sea level (5,000 feet at Cleveland, 30,000 to 40,000 feet in Pennsylvania), suggests that the basin gradually subsided as it accumulated deposits. Land masses are like great ships floating on the surface of the globe; they sink or rise as their cargoes of rock or ice are increased or decreased.

The Devonian rocks confirm the principle discovered by William Smith: that it is the fossils, not the appearance or mineral composition of rocks, that identify their place in the geologic column and in time. In any locale, the older rocks usually lie below the younger ones, as Nicolaus Stensen said, but in making comparisons between one part of the world and another (even between Wales and Devon in the case of the British rocks), it is the fossils that speak of a time shared by different (and sometimes vastly distant) places.

In the shale strata of the river gorges of northern Ohio, the immense journey of the earth itself has been silently recorded. When New England farmers first came into our area in the early nineteenth century, humanity had little idea that the earth was more than a few thousand years old or that it has truly changed, not merely endured, through time—that time and our planet, circling in the heavens, are transmuting a mortal past into an unseen and very different future. The English geologists of the 1830s discovered the history of the earth not in the rocks themselves, but in their preserved impressions of oceanic life. Biological observations infused history into geology; geology then gave a historical perspective back to biology.

The period of a rock could be read from its fossils: A major group of graptolites, for example, might be found in a rock from the Ordovician, but never from the Devonian. The idea, however, that the graptolite might be an ancient organism that later went extinct did not really come from biology itself, but was thrust on biology by the geologic fact that Ordovician rocks lay below Devonian ones, and hence must have been formed at an earlier time. Darwin was a traditional geologist (as well as theologist) when he left England in 1831 on the *Beagle,* but when he returned in 1836 he was well on his way to becoming an evolutionary biologist. The rocks of the earth record a stream of biological creation far more eternal and immanent than that portrayed in a New England Bible.

Above the black Devonian shale in the Vermilion gorge is red Mississippian shale. The change in color may denote a change from a saltwater to a freshwater environment, but the red shale is so devoid of fossils that this point cannot be certain. The red shale extends eastward past Cleveland, and in a north–south belt all the way into southern Ohio; westward it has been eroded away—not surprisingly, because it is more than 325 million years old.

Slightly younger than the red shale is the Berea Sandstone, also of Mississippian age, which lies over 200 feet thick at the South Amherst quarry. The sands that formed the Berea Sandstone may have been washed down by streams from Canada; the stone sometimes displays undulating ripples, as though its grains had just been stirred by currents of water or wind. From the drifting sands of streams and beaches 300 million years ago, compression and chemistry formed one of the most beautiful building materials in the world; many of the college and city buildings of northern Ohio have their origins in those ancient sands.

The Mississippian strata are the youngest rocks we have beneath us in the Oberlin region. It is probable that Pennsylvanian sediments accumulated here, but have been eroded away in the past 280 million years. Pennsylvanian strata cover a large belt of Ohio to the east and south of Oberlin, and come to within 20 miles of our village at places where they have escaped relatively recent erosion by streams. The Pennsylvanian period saw the rich development of land flora and fauna, with the evolution of ferns, seed ferns, conifers, and reptiles; scale trees grew to heights of 100 feet. The lush vegetation gave rise to the coal deposits of southeastern Ohio, Pennsylvania, and eastern North America, as well as those of Britain and Europe.

The earth has had sufficient terrestrial vegetation to produce coal in every period since the Devonian; important coal deposits are in the Permian (China and Australia), Jurassic (Siberia and Australia), and Cretaceous (western North America) periods. Coal does not have to come from the Carboniferous (Mississippian and Pennsylvanian) period. Nor do rocks from that period—for example the red Mississippian shales of our region, or the gorgeous red Pennsylvanian (and Permian) sandstones of the Garden of the Gods in Colorado—have to yield any coal. If the Oberlin region ever had any coal deposits, they all were washed away.

It is impossible to know exactly when the Oberlin region lost its maritime status and rose, for all time since, above the waves, but it had probably happened by the beginning of the Permian period. Sedimentary Permian rocks, indicating the continued presence of the ocean, do occur in southeastern Ohio. Waters continued to inundate the interior of North America in the western states and provinces through the Permian, Triassic, Jurassic, and Cretaceous periods.

Students coming into this tranquil midwestern college town from the bustling east and west coasts of our country are apt to remark that nothing ever happens here. Probably no one warned them that, geologically speaking, this has been a quiet place for more than 500 million years. No volcanos have ever erupted here; no clashing tectonic plates have thrust up mountains. For about 300 million years following the Precambrian, our place quietly accumulated 5,000 feet of sediments, laid down in flat layers as Stensen said they should be. For 300 million years subsequently, those layers have remained almost flat.

I have walked along the northern coast of Cornwall in England, not too many miles from where Sedgwick and Murchison studied the contorted Devonshire rocks. In that serene country, where sheep graze quietly to the edge of the cliffs, the strata speak of convulsions in the past. Layers of Carboniferous rocks stand upright, marching down from the cliffs across the sands to the sea, where ribbons of algae find anchorage on rocky keels as the tides sweep in and out. That little part of England has been rotated 90 degrees from the way that it once was created. Other layers nearby were rolled and folded as though they were modeling clay or pastry. Sometime after those Carboniferous rocks of Cornwall—or the Mississippian rocks of Oberlin— were formed, enormous geologic forces were unleashed in the world. Those forces turned and folded the strata of England, but they barely disturbed the tranquility of our region. As I look up at the shale cliffs of our Ohio gorges, I can scarcely detect the slight dip to their flaky layers.

The world underwent extraordinary changes as the Permian gave way to the Triassic period. Many marine invertebrates such as trilobites, ancient corals, spiny brachiopods, and the "rice rock" protozoan fusilids disappeared from the fossil record. Animal groups that had lived in our seas for more than 300 million years said farewell to a world no longer theirs. Giant club mosses and horsetails and huge carnivorous reptiles also became extinct; the earth may have lost 96 percent of its plant and animal species.

We do not know what caused the environmental changes that led to this Permian disaster, but stretching across 870 miles of Siberia, and bordering Lake Baikal, is an enormous band of volcanic rock (the Siberian Traps) that was formed during a relatively short period 248 million years ago. The massive volcanic activity that created those formations may also have devastated the world's atmosphere with discharges of dust and gases.

The extinction of so many species left the world wide open for the evolving descendants of the survivors. In the Triassic period, reptiles and dinosaurs became prevalent, and probably the first small mammals appeared; the world left the Paleozoic and entered the Mesozoic era, the Age of Reptiles.

Eastern North America was affected by other dramatic changes in the Permian world. An Appalachian highland was thrust up by tremendous forces affecting our continent, continuing a process that had begun in the southern Appalachians during the Pennsylvanian period. The strata of Ohio were warped slightly by those mountain-building forces, being pushed into a very gentle arch that extends across the state. Travelers from the eastern states, seeing the beautiful Appalachians end as the road leaves Pennsylvania and enters Ohio, might well wish that this flat land had felt the forces a little more fully. The slight incline formed across Ohio is called the Cincinnati Arch; its ridgeline extends from Cincinnati through

Dayton and Findlay to Toledo (and also southward from Cincinnati to Tennessee). From that line, the rock strata dip toward the southeast or, in the direction of Indiana and Michigan, toward the northwest—but so gently that the strata in the river gorges or quarries appear almost horizontal.

Lake Erie lies absolutely flat (or as flat as one can be on the surface of a spherical world), and there is no east–west change in elevation in the land along its shores. But when the Cincinnati Arch was originally formed in Permian times, its crest really did lie higher, and hence was subject to faster erosion, than the ground to either side. During the past 250 million years, the original ridge of the Cincinnati Arch has lost its Pennsylvanian, Mississippian, and Devonian strata, so that Silurian or even Ordovician rocks now lie exposed. East of the ridgeline at Sandusky, only the Pennsylvanian and Mississippian formations were lost, so that Devonian rocks lie on the surface, while at Oberlin the Pennsylvanian alone were eroded, leaving Mississippian rocks at the surface (below the glacial clay). Thus the geologic map of Ohio, because of the Cincinnati Arch, consists of a sequence of southwest–northeast stripes, depicting the strata in order, from older to younger. A similar pattern is seen in southeastern England, though with rocks of Mesozoic rather than Paleozoic age and with greater incline than in Ohio.

The formation of the Cincinnati Arch and the Appalachian Mountains was associated with truly earth-shattering events that pushed continents together and then tore them apart. Do I really know where Oberlin, Ohio, is on the surface of the earth? Looking up the numbers, I might say that it (and only it) is at 41.3 degrees north latitude and 82.2 degrees west longitude. But in geologic time, to know where a place is, is not to know where it was. Oberlin's latitude and longitude were not

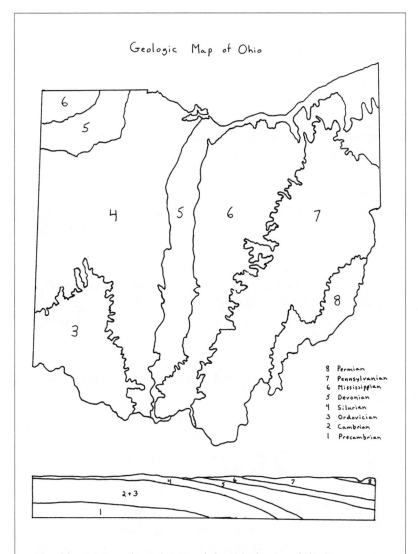

Geologic Map of Ohio

8 Permian
7 Pennsylvanian
6 Mississippian
5 Devonian
4 Silurian
3 Ordovician
2 Cambrian
1 Precambrian

Adapted from J. A. Bownocker, Geologic Map of Ohio (Columbus: State of Ohio Department of Natural Resources, Division of Geological Survey, 1981) always what they are today, and indeed what they are today is slightly different from what they will be tomorrow.

When I was a child playing with maps and globes and jigsaw puzzles, I remember thinking that South America would fit very nicely against Africa. Many other children must have noticed the same thing, but then we all grew up and learned to think less fanciful thoughts. Alfred Wegener, however, retained throughout his life his child-like imagination and enthusiasms, which took him high in balloons to study the earth's weather and far across Greenland to observe its glaciers. In 1910, his curiosity was aroused by the similarity of the coastlines of South America and Africa. The next year, he learned that many early fossils from Brazil and South Africa were so similar that paleontologists wondered whether a landbridge had once joined the two continents, but had later subsided into the South Atlantic. The little reptile Mesosaurus, for example, a mere 18 inches long, was living only in Brazil and in South Africa in Permian times, and it seemed unlikely that it was maintaining its species' integrity by swimming back and forth across the Atlantic. Wegener proposed an alternative to disappearing land bridges: that the present continents were once joined together as a single supercontinent, Pangaea, which in post-Permian times broke apart.

Geologists were skeptical for many years following the publication of Wegener's *The Origin of Continents and Oceans* in 1915, but since 1950 evidence from magnetized volcanic rocks and from the ocean floor has strongly supported the view that the continents have been drifting apart since Triassic times. As they cool, iron-containing lavas from volcanos become magnetized in the direction of the earth's magnetic field, and hence become fossilized compass needles indicating the direction of that field when they were formed. Young rocks point in the direction of the earth's present magnetic field, but older ones show considerable deviations, suggesting that since their formation they have been rotated relative to the earth's magnetic axis—perhaps by a movement of the continental mass on which they sit.

Volcanic rocks formed during the past 15 million years sometimes have their compass needles pointing exactly south instead of north, and when magnetic directions are correlated with rock age, the extraordinary view emerges that the earth in recent geologic history has reversed its magnetic field about every 300,000 to 500,000 years. On the ocean floors, these reversals of magnetism are seen in regular bands parallel to oceanic ridges, such as the Mid-Atlantic Ridge. In the center of these ridges are rifts from which basaltic rock is welling up from the earth's interior, pushing the ocean floors outward like a conveyor belt that carries the continents with it. The width of the reversal bands on the ocean floor indicates the rate at which the conveyor belt is running.

And so since Permian times, our land—all those rock strata accumulated under northern Ohio—have been going for a ride, at perhaps 1 to 2 centimeters a year, on a very slow boat toward China. We started near the northwest coast of Africa, and probably much closer to the equator than we are now, so that we have been drifting northward as well as westward. Farther south, the homeland of Mesosaurus was split in two, as South America broke away from Africa.

The new ecology created by the drifting continents and Permian extinctions saw the evolution in the Triassic of many species of small and moderate-size dinosaurs that occupied both land and sea. The Jurassic and Cretaceous periods produced the enormous animals whose bones are displayed in museums around the world. In Jurassic times, brontosaurus and stegosaurus inhabited the land, and plesiosaurus the seas; tyrannosaurus and triceratops were among the Cretaceous giants.

Northern Ohio has no rocks from these periods, and hence no fossils of dinosaurs, but it is fair to suppose that dinosaurs were running, nevertheless, through what are now our farms and villages, above the Devonian sea of placoderm fish. The greatest collection of dinosaur fossils is from the western states, especially from the Jurassic Morrison Sandstone of Colorado and Utah, but dinosaur fossils have been found worldwide. William Smith was not lucky enough to find a dinosaur fossil in the Jurassic rocks he characterized in England, but in 1826 the first English dinosaur bone was found in the limestone quarries of Stonesfield, a few miles from Smith's boyhood home in the Cotswold hills north of Oxford.

Southeast of Smith's Jurassic strata in England are the chalky hills—the North Downs and South Downs—that end, at the English Channel, in the White Cliffs of Dover, the Seven Sisters, and Beachy Head. Along those cliffs, one can pick up great chunks of the soft white chalk for which the Cretaceous period is named. In America, the thick Dakota Sandstone, which underlies the western plains and holds a large reservoir of artesian water, is from Cretaceous times.

During Cretaceous times, northern Ohio and the world saw extraordinary changes in fauna and flora, with an extensive evolution of mammals, birds, and flowering plants. While the forests of the Jurassic world were rich in conifer trees, ferns, equisetum (horsetails), and lycopodium, the temperate forests of the Cretaceous began to look increasingly like the deciduous forests of Ohio today. In spite of continental drift, Eurasia and North America were still broadly connected through Greenland, and latitudes now northern were then green and temperate, sharing east and west a similar flora of trees. Fossils from western Greenland include sycamores, oaks, magnolia, laurel, and sequoia, showing the mildness of Greenland's Cretaceous climate.

Cretaceous fossils from Maryland and the Dakota Sandstone in the West show that North America had a great many deciduous trees very similar to those we would see in the woods around Oberlin today: maple, oak, beech, poplar, sassafras, sycamore, dogwood, walnut, birch, willow, and elm. Vines of grape and of Virginia creeper were also present. The forest floor was probably covered with ferns, mosses, and fungi, but most of our familiar herbaceous wildflowers were probably missing: Woody species of flowering plants appear, in many cases, to have predated their herbaceous cousins.

It is strange to pause in our woods today to realize that a similar forest may have covered our land 65 million years ago—similar in its flora, yet so very different in its fauna. There were tall trees of beech and maple, but not the shadow of a person; no human foot would rustle the autumn leaves or break through the crust of the winter snow, until the sun had come north and gone south again millions of times. Behind the dogwood a dinosaur might stir, where today there would be deer. But the dinosaurs had few seasons left in the changing colors of this deciduous forest. Small mammals were scurrying amid the mosses and leaves. Little horses would soon greet the dawn.

The Ohio woodlands passed from the Mesozoic into the Cenozoic era, or Age of Mammals, which includes the Tertiary and our own Quaternary periods. As with the transition between the Paleozoic and the Mesozoic, there was a large extinction of animals—dinosaurs and many other forms. Presumably, environmental changes initiated this mass extinction, though the loss of one species in the web of life might become cause itself for the loss of another. Luis Alvarez has suggested that the Cretaceous–Tertiary extinction was caused by the collision of the earth with a giant meteorite, which threw tremendous clouds of dust into the atmosphere, dimming the solar light available for photosynthesis and changing climates throughout the world.

The Tertiary period saw a great evolution of mammals and of flowering plants, especially among herbaceous flowering plants, including the grasses. For perhaps the first 20 million years of this period, North America remained united in temperate latitudes with Europe on its east as well as with Asia on its west, so that the flora and fauna of the whole Northern Hemisphere developed to a large degree together. Fossils of the earliest horses from the western states, for example, are identical to those found in Britain and Europe. When the bridge to Europe was finally lost, the Bering bridge to Asia was retained, at least periodically, until about 15,000 years ago.

South America, however, drifted free of the other continents throughout much of the Tertiary, until it ran up against North America at the end of that period. Hence the plants and animals of Ohio are much more closely related to those of Asia than to those of South America. The hummingbird has reached us from South America, but the flowers that it finds here are different from those of its continent of origin. The European explorers of the New World found the plants and animals of North America much more familiar than those of South America.

Northern Ohio has no record in its rocks of the Tertiary period, or of the Mesozoic era, or of the Permian and Pennsylvanian periods before that. Our rocks are of a distant past, and we have to look to distant places to understand what our land has seen during most of the past 300 million years. Yet this place is what it is because of its connections with the whole world, and our understanding of it is rooted not only in local studies, but also in thoughts from other times and places. We might see little in our own Devonian shale had not the Jurassic limestone of England excited the thoughts of William Smith.

For the Quaternary period—the age of ice and the age of humanity—our land possesses again a direct record of its history. We will save that story for subsequent chapters.

As I pick up a flat little rock, a Devonian stone, from the gorge of the Vermilion River and skip it across the gently flowing waters, the rippling, expanding circles remind me of the words of Alfred North Whitehead: "The present moment holds within itself that whole amplitude of time, backwards and forwards, that is eternity." We live at the end of time, and also at the beginning of time. All is here that has ever been, yet nothing has ever been quite as it is here.

2 water and ice

The form of the land is shaped by water, and so is the form of life. I walked along country roads one sunny day in early March. The red-winged blackbirds, recently back from the South, were chattering in the trees above the Black River, and everywhere snow was melting along the edges of the lanes. The fields were alive with the glitter of light on patches of ice and snow, and little freshets of water ran everywhere across the land, forming their own lacy routes to the ditches beside the roads and to the meandering tributaries of Elk Creek and the Black River. The day was witness to a world of creation, as the crystals of winter liberated again the flowing waters of life. Here in a countryside of roads and agricultural fields, the wild forces of nature were everywhere awake, generating new patterns across the greening land.

The most common things of our world are also the most miraculous. All living architecture is based on water, for water is more prevalent in organisms than all other substances combined. Without this most common of substances, all the mira-

cles of living activity cease. Spirit and body instinctively sense
the crucial link between life and water. An injured animal will
seek shelter by a pool, drawn to its healing powers. A lonely
spirit may seek a flowing brook, model and metaphor of our
nature. Like a river, all life is water moving beneath the energy
of the sun, a union of the rain, the uplands, and the ocean.

The magic of water is often hidden from our view. Seldom
do we stop to think that it is the most universal of solvents,
attuned to holding the myriad chemicals of life in its gently
electric matrix. Its cohesiveness within and strength along its
surface, and its thermal stability combined with conductivity
all contribute to the organization and activity of life. And it
dwells at just the right thermal interfaces between liquid, solid,
and gas. Water, as the ancient philosophers and poets foresaw,
can hold earth and air and gentle fire within it.

When water freezes, it can suddenly reveal—to our watery
eyes and nerves and brain—the miracles it holds inside. On a
winter's morning, the first rays of the sun disperse their colors
through fractal fronds of frost on the windows. How many
hours would a craftsman labor to make such fine crystal patterns from the nighttime air? Yet from formless gas, by the mere
chill on the world, come the frost crystals on the window or the
snowflakes in the air. From simple actions and common matter,
the complex beauty of nature emerges.

That the landscape is shaped by water is evident to all who
live in a country of soft shales and clay. When I walk the
streambeds of the Vermilion River, even in the driest of summer seasons, water seeps out at many points along the high
walls of the gorge above me, and with every trickle of water, the
flakes of shale come loose and fall, sometimes forming great
conical heaps of soft stone beside the river. As winter ice along
the steep banks gives way to the rays of the springtime sun, the

shale itself seems to melt from the canyon walls, and a chorus of falling stone accompanies the songs of dripping water. In such a country, landforms give way to the forces of water before our very eyes. At the junction of Chance Creek and the Vermilion River, the little promontory on which I like to sit is greatly altered from the form it had before the torrential rains of July 4–5, 1969, when the rivers flowed 12 feet or more above their customary shallow levels and swept away much of the point. Another time will see a different world.

The sedimentary rocks beneath us were created by the forces of water on ancient landscapes, as rain and streams brought other highlands down to other lowlands. In the mud of our rivers flows the land into a new time and different form. In soft country like this, everyday observations lead easily to the idea that large floods may have helped to shape the land we see before us.

Early geologists were well instructed as children in the biblical flood, which required Noah to build an ark, and for many the account was accepted as earth history rather than human allegory. Fossils of seashells found high in the sedimentary rocks of the Alps seemed to be evidence that the biblical flood had indeed been extraordinary. If the idea seems preposterous that water once covered central Europe to a depth of 14,000 feet, how likely does it seem, from the perspective of lives on apparently solid ground, that the Alps have been raised that distance from a position previously below the ocean? Marine fossils on mountaintops show that land and water have undergone remarkable processes of one kind or another.

Other geologic features of Switzerland—and of northern Ohio—baffled nineteenth-century observers. The story that science has invented to explain them may seem far less probable than Noah's flood. The scientific verdict, almost universally accepted today, is that millions of square miles of Europe, Asia, and North America were recently covered not with 14,000 feet of water, but with an equivalent thickness of ice.

Northwest of the Alps in Switzerland and France, across the plain of Geneva, lie the Jura Mountains (for which the Jurassic period was named). High in the Jura Mountains are large rocks unlike those of the Juras, but identical to those of the Alps. They are analogous to the large rocks on Tappan Square in Oberlin: They do not belong where they are, but one does not have to go too many miles away to find where they might have come from. Then there are scratches and grooves and polish on the limestone of the Jura Mountains (as at Sandusky Bay or Kelleys Island in Lake Erie), as if giants had once carved and shined the rocks. Darwin and Sedgwick walked past such sculptured rocks in a narrow valley of North Wales as they hunted for fossils, but such features of the land seem to defy any plausible explanation.

No one living in England or in northern Ohio was likely to imagine any natural process to account for such things. But climbing in the Swiss Alps above Zermatt, I once looked down many hundreds of feet on the Monte Rosa glacier and saw its massive ice, laden with rocks, on the summer land. It is not too hard to imagine that the glacial ice might be slowly falling down the mountain.

When young Louis Agassiz went to Les Diablerets glaciers in the summer of 1836, alpine farmers and woodcutters had long known that mountain glaciers moved, that they carried rocks long distances, and that glaciers had once extended much farther down the mountains. Agassiz became convinced that mountain glaciers had once completely overrun the Jura Mountains from the Alps, carrying and dropping those erratic rocks, and carving and polishing the limestone of the Juras. From his perspective in the mountains, Agassiz could also imagine what was unimaginable to anyone in the lowlands: that ice might once have come down from the Arctic and stretched across the whole northern world. Was this a vision of intellectual genius, or an aesthetic response to the grandeur of his native mountains? Or just an expansive leap of the unfettered mind in an air made thin of oxygen? At high elevations, the mind is free to create bold thoughts. Agassiz was seized by this novel idea and soon was convincing colleagues in Britain and America that the Highlands of Scotland and Wales, the White Mountains of New Hampshire, and our own Great Lakes had been sculptured by ice. All this came to Agassiz's mind long before European science discovered that Antarctica and Greenland are even today covered by vast sheets of ice.

Two American naturalists anticipated elements of Agassiz's thought. In 1823, Ebenezer Granger argued that the grooves in a rock on Sandusky Bay must have been made by a natural force, but that the action of water would be inadequate, requiring instead some "hard body" to cut them. And in 1826, Peter Dobson suggested that boulders he unearthed in building a factory in Connecticut might once have been dragged by ice across other stone, producing the ridges and polish he found on their undersides. Both observers published their thoughts in early volumes of the *American Journal of Science*, which was little read in Europe.

The biblical flood of water did not give way to Agassiz's flood of ice without a hybrid theory. Icebergs were well known to whalers, and icebergs sometimes carried rocks. For a few years, the presence of erratic boulders was explained by a "drift theory": that icebergs floating in a flood had carried rocks to distant places. The theory died, but the term "glacial drift" remained to describe material carried by glaciers.

Glaciers could be invoked, like gods or floods, to explain almost anything, and before he was done, Agassiz was describing features of the Amazon basin (where ice sheets have never been) as the result of glacial action. Glaciers can pick up material or dump it. They can destroy hills or create them. They can scour out valleys or fill them in. They can scratch rocks or polish them.

In northern Ohio, the most pervasive glacial feature is one we seldom notice until we dig in the garden or wonder why the rain or melting snow so easily forms pools on the land: the soil over much of our land is impermeable clay. In shale country, glaciers have no difficulty in grinding up the soft rocks to make clay. In northern Ohio, this glacial till (or drift, or ground moraine) is often 20 to 50 feet deep; beneath parts of the Cuyahoga River, its depth reaches 500 feet.

Such extraordinary scraping and filling has flattened the topography of our region and divided the state of Ohio into two parts that can be clearly seen on a road map: the flat, glaciated part in the north and west, where so many of the roads run straight east and west or north and south; and the unglaciated part, where the highways twist around the hills. In some of the glaciated region, the flatness is relieved by glacial end moraines—ridges or hummocks formed where the melting edge of the glacier dumped especially large amounts of material. A series of end moraines lies across southern Lorain and northern Ashland counties; Highways 58 and 89 pass over them, like a roller coaster, north of Interstate 71.

The glacial scraping and filling also changed the pattern of streams across Ohio, even in southern parts of the state beyond the limit of the ice. In the early 1890s, a young geology professor at Denison University, William George Tight, who recently had studied at Harvard under the eminent William Morris

Davis, looked at the landscape of southern Ohio with an eye for possible changes in the drainage patterns. He soon concluded that streams now flowing south into the Ohio River had once flowed north and west across Ohio: The preglacial Muskingum River was joined by ancient forms of the Hocking, Scioto, and Miami rivers to form a river that flowed west across Indiana and Illinois to the Mississippi River. Tracing the pattern into West Virginia, William Tight and Frank Leverett linked the Ohio drainage with that of the present Kanawha River, which they thought once flowed through the Teays Valley west of St. Albans. Eventually, the entire preglacial river system was named the Teays. Its tributaries reached across Virginia to the mountains of North Carolina.

The glacier in the north blocked the outlet of the Teays and filled parts of its valley with 200 to 600 feet of drift. Waters melting off the glacier, along with those coming from the south, formed a glacial lake (now called Lake Tight) that spilled off toward the west, contributing to the formation of the modern Ohio River. Glacial meltwaters also established new southward-flowing tributaries of the Ohio.

East and north of Oberlin, the glacial till gives way to soils deposited from a postglacial lake whose waters were about 200 feet higher than those of the present Lake Erie. The shores of this ancient Lake Maumee, and the division between glacial till and postglacial lake sediments, are defined by a sandy beach ridge along old Route 10 (Butternut Ridge Road), West Ridge Road, and the Elyria–Milan road (Route 113). The beach ridge is interrupted only by the erosion of modern streams, as by the east and west branches of the Black River. North of the Lake Maumee beach ridge are two further substantial beach ridges, approximately 150 and 100 feet above the present Lake Erie, which were formed by lower stages of the postglacial waters.

Known as Middle Ridge and North Ridge, they were the shore-lines of ancient Lakes Whittlesey and Warren, respectively.

The glacial till, the lake sediments, and the beach ridges are the most extended signatures left by the glacier on the land of northern Ohio. We have no fjords, no mountain cirques, no U-shaped valleys, no hanging waterfalls (like Taughannock in the Finger Lakes), or even any significant drumlins (little hills deposited by glaciers). The glacier in our region tended to level things out instead of creating grand new formations. An exception was the gouging of a shallow basin that created Lake Erie from what would otherwise be a river. Camden Bog and other "kettle lakes" are also gifts of the glacier. Hummocky kames and a few crooked ridges of eskers to our south and east were generated where the leading edge of the glacier melted.

The giant grooves on the limestone of Kelleys Island, and the granite rocks scattered across our glacial till, are still the most dramatic evidence of the movement of ice across this land. The glacial till is a mixture not only of clays, but of sands and pebbles and larger stones and boulders. The larger the particles, the farther they were transported by the glacier: the clays traveling only a few miles; the boulders, a much greater distance.

Granitic stones and boulders are numerous in the bed of Chance Creek, where they have fallen in from the land above as the stream eroded through the glacial till and the Mississippian and Devonian shale. Large glacial boulders sit precariously on a steep knob, some 65 feet above the level of the water, at the junction of the east and west branches of the Black River in Elyria—ready to fall into the river when floodwaters undercut the shale beneath them. In such a quiet scene, unspoken but compelling evidence exists that the glacier and its rocks were there before the streams began to cut their gorges.

As I walk the streambeds of this gentle land, I feel that I am part of three vastly different dimensions of time, all united by some cosmic logarithmic scale condensing them into a single moment and into one landscape of rock and ice and water. The shale and sandstone strata of the gorge hold the shadows of the Paleozoic, 300 million years ago. The glacial boulders speak of an era 10,000 times more recent (yet still unfathomably before my time), when Pleistocene ice dropped its till. And the waters flowing through the gorge hold the secrets of postglacial millennia when the ice retreated, the lake formed, and the beautiful gorges were cut. Into such eternities, people of our own unsettled time have dropped the discarded tire or washing machine, the glass bottle and the plastic cup—like sweepings of the street cast into the aisles of an ancient cathedral.

Great ice sheets spread across the northern continents several times in the past 2.5 million years, retreating during warmer interglacial periods. In some places in the northern world, a record is clearly written: when, for example, a peat bed containing the bones of a hippopotamus and the pollen of warmth-requiring plants lies between two layers of glacial drift, indicating a warm interglacial between two periods of ice. In Ohio the records are more subtle, and in our own region the last advance of ice, called the Wisconsin Glaciation, largely destroyed whatever the earlier glaciers did.

On a planet of drifting and colliding continents, evolving life, erosion, and mountain building, no time is like any other, but our own Quaternary period is unique in two respects: in the evolution of humanity, and in the glaciation of the northern continents. Parts of the Southern Hemisphere were glaciated in Carboniferous or Permian times, before Pangaea had broken up, but the northern world had never before seen continental glaciers. Agassiz perceived in nature events as unique as Noah's flood.

Masses of ice can form and grow only in places where temperatures are below freezing for much of the year, which means at either high elevations or high latitudes. Antarctica has both high elevation and high southern latitude, and its resulting ice sheet would reach out farther than it does were it not surrounded by ocean currents and raging storms that break off and melt its flowing ice. Ice sheets of great thickness form only on land, not on the sea. In the Northern Hemisphere, the pole is occupied by the Arctic Ocean, so that glaciers, if they are to form, must have their centers on land some distance from the pole. To understand why glaciers might form where they never had before, it is necessary to speculate on factors that might change the temperatures and snowfalls of the far northern lands.

Across the shallow pools of the Vermilion River, I watch the autumn leaves caught by the complex currents, swept inward to catch the shaly shore, then back out toward the central flow again, orbiting around and about until, by some unseen variation in the flow, they are released to find another pool and another pattern of motion. For all its grandeur, the earth is like a leaf swirled in the eddies of galactic time. Science was born partly from the perception of the orderly and predictable motions of the heavens relative to the earth: that night follows day, and winter follows summer. But modern science has learned that the linkage of simple forces and motions can lead to complexities that defy exact prediction. Such is the case with weather and climate.

We seldom give an appreciative thought to the unique arrangements that make life possible on the earth. The intensity of sunlight, and the amount of energy received from it, varies inversely with the square of the distance from the sun; if the earth were twice as far from the sun, it would receive only one-quarter as much solar energy—and even the warmest parts of the earth would be colder than the Arctic and Antarctic regions. If the earth did not turn on its axis, half the earth would receive twice the energy and the other half would receive none. If the earth did not revolve around the sun, with its axis of rotation tilted to the plane of its orbit, there would be no seasons—no yearly averaging of the heat of summer with the cold of winter, no changing lengths of days and nights. On a spherical earth, the regions that face the sun perpendicularly receive far more energy than do those that face the sun less directly, but this advantage is shared across a latitudinal band of 47 degrees by the oscillations in the tilt as the earth revolves around the sun. The biomes of the world depend on such devices. The leaves that swirl before me in the Vermilion River would be other leaves if any of these arrangements were slightly different.

The earth as a whole must lose as much heat as it gains in an average day, by reradiating energy to outer space; if it did not, we would get ever warmer or ever cooler. But equatorial regions receive directly from the sun, on average, 2.5 times as much heat per acre as do polar regions. The equatorial regions receive more energy than they can reradiate to outer space, while the polar regions lose more heat to outer space than they receive directly from the sun. The earth becomes a heat-transfer engine to make up for these imbalances—removing heat from the equatorial regions and carrying it, by wind and water, toward the polar regions.

Throughout most of the earth's history, there has been no serious barrier to the transfer of heat from the tropics to the poles. But with the drifting of the northern continents in Tertiary times, the Arctic Ocean became encircled by Eurasia and North America, largely cut off from oceanic currents originating in the tropics. Likewise, the South Pole was insulated from equatorial warmth by the drifting of the large Antarctic continent to its present station. On land masses within both the Arctic and Antarctic circles, glaciers could begin to form because they were insulated from tropical warmth. The glaciers of Antarctica were (and are) limited by a surrounding ocean, but those of the Arctic could spread down across a large part of the world's land mass.

Nevertheless, the glaciers from the north have not come and stayed (as in Antarctica); they have come and gone in oscillations of glacial and interglacial periods. When the earth was set spinning and circling the sun, it was given a slight wobble of both axis and orbit. The elliptical orbit around the sun varies, from slightly more elongate to slightly less, with a periodicity of 96,000 years. The angle of the axis relative to the plane of its orbit oscillates from 21.5 to 24.5 degrees (it is now 23.5 degrees), with a period of 41,000 years. And the direction that the axis is pointing wobbles (for any particular spot along the earth's orbit), with a period of 21,000 years.

At the present time, winter in the north occurs when the earth is closest to the sun in its orbit. In 10,500 years, it will occur when we are farthest from the sun, and in 21,000 years winter, defined and caused by the tilt of the axis, will return to where it is now. Winter in the Northern Hemisphere is moderated (as now) or aggravated, depending on whether it occurs when the earth is close to or far from the sun. The change in the angle of the axis tilt influences the difference between summer and winter, a greater angle causing greater seasonal change. All these factors affect the distribution of heat received by the earth, and especially the contrast of seasons at high latitudes. A relatively low contrast between summer and winter favors the growth of glaciers: Slightly warmer winters at high latitude favor higher precipitation as snowfall, with the resulting formation of ice, and the cooler summers reduce the melting of that winter ice.

Glaciers from the north may grow and retreat as the result of the interactions of these various factors. All the beauty and history of our northern world, including the evolution and history of humanity itself, has probably been shaped in part by three wobbles in our planetary top.

As I look out at Oak Point on the vast waters of Lake Erie, the expanse seems oceanic. It is difficult to imagine that elsewhere in the world such a lake is dwarfed by expanses of ice. How can it be that the ice covering Antarctica and Greenland contains more than 200 times the water of all the freshwater lakes in the world? Of the world's total water, 97.2 percent is held by the oceans, and 2.15 percent by the ice of Greenland and Antarctica; most of the remaining 0.65 percent is in the groundwater of the land rather than in its streams, lakes, and inland seas. Only about .0001 percent is in the atmosphere as water vapor, however humid a summer's day in Ohio may be. Ice is the main contender with the oceans for the world's water, and there has been a tug-of-war between ocean and ice for the past 2 million years. When the ice wins, we have a glacial period; when the ocean wins, an interglacial period.

Each year, the earth's land and water surfaces lose by evaporation to the atmosphere 95,000 cubic miles of water, but fortunately we gain it all back as rain or snow. This colossal transfer of water is only a minute part (0.029 percent) of the world's water reservoir or a small fraction (1.36 percent) of the water currently locked in ice sheets. Only 25 percent of the yearly precipitation falls on the continents (the rest on the oceans), so that even if much of the precipitation is snow, the balance between ice and ocean can only be slowly changed.

Over thousands of years, however, ice will accumulate if the winter snowfall exceeds the summer melt. In the most recent glaciation of North America, the heaviest accumulation of ice was centered near Hudson Bay. This ice generated the Laurentide Glacier, which reached Ohio to the south and the Atlantic on the east, and, spreading northwest across the Dakotas and Montana, approached the Canadian Rockies. A second glacier, the Cordilleran, spread from the Coastal Ranges and Rockies of Canada, approaching the Laurentide Glacier near the line of the Mackenzie River. In Eurasia, the largest glacier was the Scandinavian Glacier, which coalesced with glaciers coming out of the highlands of Britain, but did not quite reach those coming out of the Alps. Much of eastern Siberia and Alaska escaped glaciation in spite of their cold temperatures.

The present-day ice sheets of Greenland and Antarctica give us a glimpse of the glaciers of 20,000 years ago. Drillings (begun in 1929 by Alfred Wegener in Greenland) show that the ice has a maximum thickness of 11,000 feet in Greenland and 14,000 feet in Antarctica. Such dimensions are staggering. The ice of a continental glacier can be ten times taller than the skyscrapers of New York or Chicago, and as tall as the Rockies or the Alps. In eastern North America, the glacier rode right over the tops of the Adirondacks, Mount Washington in New Hampshire, and Katahdin in Maine. Some of the mountains overrun in Quebec and Labrador are nearly 6,000 feet from top to bottom, so the ice must have been at least that thick in those places. How thick it was in northern Ohio we cannot know, but we can suppose that it was many times taller than the highest buildings of Cleveland.

All this ice got its water ultimately from the oceans, causing the sea level to fall by as much as 300 feet. As a result, broad areas that are covered by the sea today became dry land during the glaciations: the North Sea and Baltic Sea, the Irish Sea and English Channel, Hudson Bay and the Bering Strait. Most of the land thus exposed, except at the Bering Strait and the English Channel, was overridden by glacial ice. Seacoasts all around the world, including the Atlantic seacoast of North America, extended farther out to sea. Great Britain and Ireland became an integral part of Europe, and, most important of all, Eurasia and North America became one gigantic continent, united by a broad expanse of unglaciated land, Beringia, connecting Siberia and Alaska. The dry land of this huge continent was, however, split in two by the ice sheets of the Cordilleran and Laurentide glaciers.

In northern Eurasia and in Alaska, these great ice sheets became part of peoples' lives. Unfortunately, writing had not yet been invented, so no impressions were being recorded in a journal. Neither have we yet any pictorial representations. The artists of the time, as in the Lascaux cave in France, either lived too far from the ice or decided that there was no point in trying to represent (or to influence by magical powers) a mile-high ocean of ice.

In northern Ohio, probably no one saw the glacier until it was in slow retreat toward Canada. By 48,000 B.C., people from Siberia had occupied parts of Alaska. But we do not yet know when those people were able to find their way to the remainder of the Americas south of the ice—by migrations along the Pacific coast, or through an ice-free corridor between the Laurentide and Cordilleran glaciers. By 9500 B.C., people were living south of the ice, from the Pacific to the Atlantic, but no proof is yet available that they were here before that date. In 9500 B.C., the ice had melted from Ohio, and just north of Oberlin were the high shores of Lake Maumee.

The ice began to retreat from North America and Eurasia

about 16,000 B.C. By 13,500 B.C., the ice had surrendered enough water back to the oceans that the Bering Strait was formed again, separating North America from Asia. But the water melting off the glacier in North America had difficulty in finding its way to the ocean, for the ice had greatly altered the erosional topography of the land. The ice tore up enough of the soft shale of northern Ohio and southern Ontario to create the shallow basins of Lake Erie, but at the same time filled in with till many of the previous streams. Across a relatively flat Midwest, much of the water had no way to go. About 12,000 B.C., waters south of the glacier had formed Lake Maumee in northern Ohio and Lake Chicago at the southern end of today's Lake Michigan. To the west of the present Great Lakes, the enormous Lake Agassiz formed about 10,000 B.C., as did the huge Lake McConnell farther northwest. To our east, the rising ocean encroached on a land held down by the weight of its ice, and seawater came up the St. Lawrence valley to form the Champlain Sea.

The waters of Lake Maumee ran up against a wall of ice to the north and east; a spectacular sight it must have been, with icebergs breaking off into the lake. The eastern part of present Lake Erie was still covered with ice, so no drainage was possible through a Niagara River. The waters rose until they found an outlet to the southwest through the Maumee River (which then ran out of the lake instead of into it), into the Wabash, the Ohio, and the Mississippi rivers.

As the ice retreated farther to the north, a southern outlet became exposed across the lower peninsula of Michigan. The waters of Lake Maumee fell about 50 feet to become Lake Whittlesey, which drained through the Grand River valley of Michigan to join Lake Chicago, flowing out the Chicago River into the Illinois River and the Mississippi. As the Grand River cut into the land, the level of Lake Whittlesey fell to that of Lake Warren. Ice still occupied the Niagara region and the bed of the future Lake Ontario.

Lake Warren may have created a discharge path across upstate New York, via the Mohawk and Hudson rivers, before it finally found a free Niagara exit to the early Lake Ontario (Lake Iroquois). Lake Iroquois, in turn, found an exit into the Mohawk and Hudson rivers until the ice melted away from its outlet into the St. Lawrence.

During all this time, Chance Creek and the Vermilion and Black rivers were starting to cut their way through the glacial till and the Mississippian and Devonian shale, carrying their water into Lakes Maumee, Whittlesey, and Warren, and cutting deeper into the land as the level of the lake dropped. It is fun to think how many different paths the water from Chance Creek has had to take in finding its way to the ocean—either to the Gulf of Mexico or to the North Atlantic. The continental divide separating the watersheds of the Gulf and the Atlantic shifted with the melting of the ice, and the northernmost parts of Ohio were first on one side of that divide, and then on the other. If a Native American child had put a toy canoe into Chance Creek in the days of Lake Maumee or even Lake Warren, it might have been carried to the Gulf of Mexico via the Mississippi. Later it might have gone to the Atlantic via the Hudson River or, more recently, the St. Lawrence.

The effects of the postglacial lakes can be seen in the sandy beach ridges south of Lake Erie and in subtle features of our river gorges. The Vermilion River has cut an almost uniform incline for miles upstream from the present Lake Erie and must be cutting downward now only very slowly, since its incline is so gentle. As one walks upstream, the floodplains by the meandering river in the gorge occupy only slightly greater elevations than do the floodplains downstream. When the water level of the Erie basin was higher in postglacial times, the Vermilion River entered the lake at a higher elevation (at about 780, 730, and 680 feet, respectively, for Lakes Maumee, Whittlesey, and Warren), and thus could not cut as deep a gorge as it has today. The river entered Lake Maumee at the beginning of Swift's Hollow, so that its course from Swift's Hollow to the present Lake Erie has been cut since the retreat of Lake Maumee. When Lake Maumee fell back to Lake Whittlesey, the river turned eastward to find a continuing course, and entered Lake Whittlesey north of Chance Creek but south of present Mill Hollow. When Lake Whittlesey retreated to Lake Warren, the river cut through to Mill Hollow and beyond to about the Franks Site (Chapter 3), where it entered Lake Warren. When Lake Warren retreated, the river began to cut its final section to the present lake. At each retreat of a lake level, the river not only cut a new course downstream, but deepened its cutting on the previous sections upstream.

At each postglacial lake stage, the Vermilion River would have had a sequence of floodplains gradually increasing in elevation upstream, but all of them at higher elevations than those today. Almost all these ancient floodplains have been destroyed as the lake level fell to a lower stage and the river cut deeper. Between Swift's Hollow and Chance Creek, however, there appear to be two pieces of old floodplain that escaped the cutting of the river. One of these, the Leimbach Site (Chapter 3), became very important to the human history of this region. The flat portion of the Leimbach Site is at 700 to 710 feet in elevation, roughly midway between the top and the bottom of the Vermilion gorge. This flat area was probably once a floodplain—when the river was emptying into Lake Warren at 680 feet. When Lake Warren retreated to Lake Erie, the Vermilion

River began to cut the nearly vertical walls of the present-day Leimbach Site. Most of this cutting had probably already been achieved by the time (about 500 B.C.) that Native Americans are known to have occupied the site.

Just downstream from Chance Creek are the remains of a meander channel no longer used by the Vermilion River, a horseshoe that has been cut off by a new and shorter channel. There is a swampy pond where the river used to go and an isolated little hill that may be the remains of a floodplain from the Lake Warren stage.

The width and shape of our river gorges today also reflect the influence of those ancient postglacial lakes. A river cutting through a steep incline tends to form a straighter path than one flowing across a gentle slope. In Ithaca and the Finger Lakes region of New York, the streams cut relatively straight through the steep hills, forming narrow gorges with vertical walls close to the water on both sides. In northern Ohio, the postglacial topography presents only a gentle slope into the Lake Erie basin, and the streams form a meandering course. The postglacial lakes kept that incline especially gentle during the early stages of the streams and, in falling back in small increments along a shallow lake bottom, maintained it so throughout the cutting of the gorges. As a result, the gorges were cut by the sweeping (and changing) meanders of gentle streams and are therefore relatively wide. Today we can enjoy a double meander: a serpentine river flowing from side to side within the broader sinuosity of the gorge. As one walks the riverbanks, a vertical shale cliff is on one side and a broad floodplain is on the other, oscillating left to right in intriguing patterns as one proceeds. Mystery and expectation fill the scene ahead, for the river is always flowing out of sight, and cliffs appear and disappear as they meet and deflect its waters.

After Lake Warren became Lake Erie, with an outlet via the Niagara River, and the glacier melted far northward into Canada, an enormous lake formed north of the Great Lakes: Lake Ojibway, which emptied by the Ottawa River into the St. Lawrence. About 6300 B.C., Lake Ojibway found a new outlet to the sea, and emptied with a huge flood into Hudson Bay.

As the ice retreated, the climate warmed in Ohio and elsewhere in North America. It became warmer, in fact, than it is today. This "climatic optimum" occurred about 5000 B.C. in northern Ohio; it occurred much earlier in the Far West (about 8000 B.C. in Alaska) and much later in regions in the East and North (2000 B.C. in Labrador). Fossils indicate that during the climatic optimum many plants and animals ranged farther north, and higher into the mountains, than they do today. White pine, for example, grew 100 miles farther north of Lake Superior and 1,000 feet above its present limit in the White Mountains of New Hampshire. It is possible that part of this warm period was also considerably drier in some areas. On the Canadian side of Lake Erie, at Point Pelee, prickly pear cactus survives today. It may be a relic from a dry period in the climatic optimum, when the sandy soils of our region were home to such desert plants.

The last of the Laurentide Glacier melted east of Hudson Bay about 4500 B.C.—around the time of the warmest climate in the southern Great Lakes. The oceans reached approximately their present level. In Britain, the English Channel was formed about 5500 B.C. But as the seas rose, the lands, relieved of their load of ice, also tended to rise. The encroachment of the sea on the St. Lawrence and Lake Champlain region ended about 8000 B.C., but certain seaside plants that thrive in sandy or stony beaches (whether next to salt water or freshwater) had already established themselves along inland shores, giving our

Great Lakes a further maritime quality. When I find the sea rocket or seaside spurge growing in the sand and stones of the Lake Erie beaches, I think of those ancient postglacial times that brought the ocean so close to the southern Great Lakes.

In the depth of winter, our little Chance Creek, or even the Vermilion River, may freeze across the surface, its waters running silently below. I have skied on a winter's evening through the snowy turns and sweeps of the Vermilion gorge, beside ice falls hanging from the shale cliffs; in the moonlight, the frozen river reflects a memory of ages past, when ice lay everywhere on this land. As spring begins to dissolve these frozen scenes, ice crystals still form in shallow pools at the river's edge, among scatterings of Canadian granitic rocks, dropped during our glacial nights. The icy fern-like, feathery patterns speak of the immense creative power within the simplest elements of nature—a universal presence that shapes the long history of this and every other land, uniting us at any time or place with the elements of all the stars.

3 people

Autumn is a time of remembering. The dried corn and pumpkins in the field remember the summer rains, and as evenings lengthen, humanity feels closer to its past. When the water trickles low in the Vermilion River, I walk the dry, pebbly streambeds, thinking of the early Americans who harvested their crops on the floodplain in Swift's Hollow. Around each twist of the river's gorge, I hope to catch up with the paddlers of the past or find some stony implement that they have dropped.

It is too seldom, across our American land, that we feel the presence of those who loved this place centuries or millennia before our time. They have left so little for us to see, and our historical sensibilities are still too fractured by feelings of "they" and "we." In northern Ohio, the discontinuity was accentuated by the near disappearance of the Native Americans before the settlers from New England arrived.

In England, I like to walk the public footpaths from one village to another. Common rights to walk carefully around, or even through, another man's fields and woodlands have been handed down since Celtic days, when land was public for those who used it. Nature and humanity are everywhere woven together in the English countryside. Narrow lanes worn down through the soil by centuries of footsteps and farm cart wheels, luxuriant old hedges to either side; earthy undulations of pasture ridges and furrows, left by medieval plows; lichens and mosses long undisturbed on country churchyard walls; and small groves of trees against the distant sky, atop ancient burial mounds: All quietly remind us of the human toil, hope, grief, and love that shaped, with nature, the fabric of the land. On the windswept downs of southern England, among the giant rocks of Avebury and the grassy top of Windmill Hill, one can almost see the Neolithic people herding their cattle, harvesting their grain, and offering prayers to the earth, the sun and moon, and the space and spirits of their world.

I drove out on West Ridge Road the other day to visit with a man who has for eighty years explored the ancient human origins of the northern Ohio land. Raymond Vietzen grew up on the old beach ridge of Lake Maumee, and the stone relics he picked up in the fields of his family's farm showed him at an early age that humanity had inhabited his land for many hundreds of years. Behind his house he has built, as a memorial to those people, a rustic museum filled with Native artifacts. We sat on the little back porch of his house, facing the museum, while he recounted the adventures of excavating Native American sites along the Vermilion and Black rivers. The books he has written are full of interesting findings, described in appreciative and reverent tones. We would all do well to remember, as he does, the generality of our human ancestry and to rejoice that our roots are in a land that has long been loved.

The peoples who cherished the New World for eleven millennia before the *Santa María* arrived from Spain could have had no name for their collective whole, for they were spread across thousands of miles of two continents, as well as thousands of years of history. Nor did these early Americans know that they had discovered a new world, for there was no ocean to cross from Siberia, nothing to mark the end of one continent and the beginning of another. Animals and plants had known the two continents as one throughout most of the Tertiary period, until the sea flooded the Bering Strait about 15 million years ago. The Bering region has subsequently alternated between being flooded or being dry, as sea and land levels have changed, and animals have migrated when the land was open.

Early camels, horses, and lynxes migrated to Asia from probable origins in North America, but most of the traffic has been the other way, from the larger continent to the smaller. Long before the Wisconsin Glaciation, the mastodon, woolly mammoth, elk, caribou (reindeer), musk ox, bison, and black bear—grand sources of food for people—entered North America, along with the wolf, red fox, wolverine, and weasel. But people, following the animals, did not reach the Bering region and Alaska until the Wisconsin Glaciation, about 50,000 years ago. While the animals dispersed to the lower part of North America during warm interglacial periods preceding the Wisconsin Glaciation, the people arriving in Alaska during the Wisconsin Glaciation were probably blocked by ice from the rest of the New World. The food for big-game hunters was there, but for a long time it could not be reached.

By 9500 B.C. (some think it was earlier), people had burst into the silent kingdom of big animals south of the ice. How disconcerting it must have been for those animals. For tens of thousands of years, camels, horses, woolly mammoths, mastodons, elk, and bison had roamed through the fields and forests. While they had to be wary of saber-toothed tigers, cougars, huge short-faced bears, and dire wolves, they had never known the cunning creature with two legs who suddenly appeared among them with his long arms and spear. Between 10,000 and 7000 B.C., many of these large mammals were extinguished. Some, like the American camel and horse, had close cousins that survived in Eurasia. Others, such as the mammoth and mastodon, went extinct in Eurasia as well. We will never know the degree to which our ancestors are responsible for the extinction of these large animals, but that they hunted and killed them is almost certain. At a site near Folsom, New Mexico, in 1926 a flint spear point was found lodged between the rib bones of an extinct bison. Folsom-type "fluted" points, which have a central groove chipped out to aid in attaching a wood shank for a spear, date from 9000 to 8000 B.C. Slightly older Clovis points (first found near Clovis, New Mexico) have been found with mammoth bones at many sites dating from 9500 to 9000 B.C.

The Americans who made Clovis and Folsom points were Stone Age people. Indeed, they were of the Old Stone Age, or Paleolithic, and so can be called the Paleo-Americans. But their technology and audacity seem to me as amazing as the moon landings of the twentieth century. How can anyone feed his family by fashioning sticks and stones into projectiles for killing enormous elephants? But so they did, and so successfully that their families spread from the Pacific to the Atlantic, and throughout South America as well. Agriculture had not yet been invented; these earliest Americans were big-game hunters.

The fluted point has been found at many places around the Great Lakes. At least 500 fluted points have been found in Ohio, and so have the bones of many mastodons. People were probably in the Oberlin area by 9000 B.C.; they may have missed the high waters of Lakes Maumee and Whittlesey, but they very likely saw the late stages of Lake Warren before it fell back to the level of Lake Erie.

Unfortunately, the big game soon gave out. The hunting may have been just a little too easy—if killing mastodons and mammoths with stone-pointed spears can be considered easy. The large animals of North America, isolated from people for so long, may have been behaviorally ill-adapted for avoiding the hunt. Similar species of camels and horses, for example, survived in Eurasia where they had a long and continuous experience of people, but were eliminated in North America where that experience had been missing. The climate was changing dramatically across the northern world when these extinctions occurred, so that other environmental factors may have been more important than people. Strong winds (generated by the cold glaciers) blowing across bare soils probably created dust storms that would have dwarfed those of the Dust Bowl of the 1930s. With the warm, dry conditions of the climatic optimum, grasslands probably provided less food, and water became harder to find, so that the populations of large herbivores diminished and survivors had to stay close to limited water courses, where they were easily hunted. Whatever the reasons, 35 to 40 species of large mammals became extinct in 3,000 years, compared with 9 to 10 species in the preceding 70,000 years.

In northern Ohio and the Great Lakes generally, the biggest game was the mastodon. The mammoth lived on the tundra and in open grasslands, and the Great Lakes area rapidly became forested as the glacial ice melted back. Tundra develops where permafrost in the ground prevents the establishment of trees, and grasslands succeed where rainfall is sparse or where other factors discourage woody plants. When the ice melted in our area, the soils thawed and the land scarcely went through any tundra stage before the trees returned. The mastodon, browsing on tamarack and other trees, was at home in the woodlands, but the grazing mammoth needed open places. A great many mastodon bones, but relatively few mammoth remains, have been found in the Great Lakes area.

As the mastodons and mammoths disappeared, the hunters turned to somewhat smaller game. The large *Bison antiquus*, with a shoulder height in excess of 7 feet, went extinct; but the smaller, modern species survived and became a primary target on the grasslands, where they were sometimes herded into natural canyons or corrals that the people built. In the woodlands, the caribou gradually replaced the mastodon as a major source of food. The fluted Folsom points of the big-game hunters began to give way to nonfluted, long narrow points (lanceolate points) and points with stems for the attachment of shanks. The people who made these points are called Plano-Americans because so many of their sites are in the Great Plains, but they also occupied the Great Lakes region, where they are called the Aqua-Plano people because so many of their sites are on the beach ridges of the postglacial lakes. All the Americans from Clovis to Aqua-Plano times (9500–5500 B.C.) are grouped together as the Paleo-Americans to emphasize their common dependence on hunting large animals.

Before the Paleo-American time was over, however, great changes in human life were occurring. People began to look for different ways of making a living as the big animals gave out, just as happens today when a major industry begins to fail. There are signs of greater reliance on plants for food, with stones used for grinding wild plants. In a rock shelter overlooking the Mississippi River in Illinois, a domesticated dog lived in

6000 B.C. People were nurturing, close to home, individual plants and animals of their environment.

Agriculture was invented in America about the same time that it spread across northern Europe to reach England. The first varieties of maize, originating from the wild teosinte grass, may have been grown in southern Mexico as early as 4000 B.C. The earliest forms of maize produced very small ears, and many centuries of artificial selection were necessary before maize could become a primary source of food.

The New World presented greater geographic difficulties than the Old World for the spread of agricultural crops, for its north-south orientation meant that dissemination had to be through greatly changing climatic zones. The spread of maize northward was impeded by very dry regions and increasingly cool climates. In contrast to wheat and the other cereals of Eurasia, maize had its origin in a warm climate and required slow genetic changes, by continual selection, for its successful growth in cooler regions. Unlike the Eurasian grains, maize cannot compete with native weeds without human cultivation and will die out in a field within a year or two. The spread of maize northward was therefore slow: It reached the southwestern United States about 1200 B.C. and began to be grown in Ohio about the first century A.D. The introduction of maize into the Ohio region was contemporaneous with the development there of the extraordinary Hopewell culture.

In the Great Lakes area, and the eastern woodlands more generally, the period 5500 to 500 B.C. is known as the Archaic. During these five millennia, great diversification occurred in the ways of life of Americans living in different areas. The huge mammals were gone, and with them the relative uniformity of the Paleo-Americans. The Archaic peoples of the Great Lakes area took up fishing in a big way. They made fishhooks from bone and antler, fish spears, nets and notched stones as sinkers for the nets, and probably traps and weirs for controlling water flow. They hunted small animals with snares and darts. They made baskets for collecting and storing foods, and mortar and pestle–like stones for grinding plant foods.

Until recently, it was thought that agriculture did not emerge in the eastern United States until the end of the Archaic period, when maize, squash, and beans were introduced from Mexico and Central America. There is now evidence, however, that agriculture developed in the eastern states independently of middle America, perhaps as early as 2000 B.C., with the domestication of at least four native plants: goosefoot, sunflower, marsh elder, and an eastern species of gourd or squash. Goosefoot is a weedy plant that quickly invades disturbed soils. I created a new path in my garden last summer, and this summer pulled up scores of goosefoot plants that had invaded it. When I once built a log cabin in the Minnesota woods, goosefoot was the first plant to spring up in the sandy soils around the cabin the next summer. The goosefoot's plentiful seeds are nutritious, as are its young leaves. It seems natural that hunter-gatherers might first harvest it from disturbed areas around their camps and later begin to plant its seeds.

Stores of goosefoot seeds have been found in caves in the eastern states. At Ash Cave in central Ohio, a large cache of goosefoot seeds was found in 1876. A sample of the seeds was sent to Asa Gray at Harvard for identification and preservation. Bruce Smith of the Smithsonian Institution recently examined the seeds with electron microscopy and found that their seed coats were much thinner than those of wild goosefoot. Thin seed coats favor early sprouting in the spring; under conditions of domestication, plants coming from and producing such seeds might well be favored over those producing thicker coats.

Such a marked difference between ancient stored seeds and those of present-day wild plants strongly suggests that the Native Americans were not just gathering goosefoot seeds, but also farming with them. Likewise, differences between recovered and wild-type seeds of sunflowers, marsh elders, and gourds suggest domestication of those plants also during the Archaic period.

Pottery was absent in the Great Lakes region during the Archaic, but a limited use of copper began here earlier than anywhere else in the world. Five thousand years before Charles Martin Hall figured out, in a back shed in Oberlin, how to extract aluminum metal from bauxite, Americans had discovered metallic copper in the rocks of upper Michigan and had begun to fashion spear points and fishhooks from it.

Much of what we know of our Archaic ancestors comes from burial sites. The response to death is a universal one, whether implements are made of stone, copper, or plastic. On a high ridge above the stone avenues and circles of Avebury in England is the West Kennet long barrow, a burial chamber constructed of massive stones and earth. It is not hard to understand why it is where it is. I have stood on that ridge on a blustery day and felt my own spirit carried with the wind and clouds across a vast landscape joining earth and sky. Bones and skulls and precious artifacts were buried at West Kennet at the same time as our Archaic people buried their dead in their own special places, as on the ridges of the glacial lakes or those of the glacial kames. In both England and America, burials were sometimes ornamented with red ocher pigment, perhaps in a shared belief that redness, like blood, is an essential part of human life.

From the high ground to the low, the Archaic people knew every inch of our land. They traveled up and down the Vermilion River, and must have been very familiar with each turn that it makes. The headwaters of the Vermilion are at Savannah Bog in Ashland County, a few miles southwest of Oberlin. In 1976, a large dugout canoe was discovered in Savannah Bog. It had been hewn from a large white oak log in 1600 B.C. It is nearly 23 feet long and weighs, when reasonably dry, over 850 pounds—not an easy craft to portage from the Vermilion River to the Mohican. The Native Americans often buried their canoes underwater for the winter, a natural cold storage, and the acidity of the bog would preserve the craft even better. Someone must have moved on before spring, and the location of the canoe was forgotten, but the bog was patient in preserving its prize. It now resides in the Cleveland Museum of Natural History—the most spectacular relic we have of these people who paddled our streams so long ago.

About 500 B.C., the Ohio country entered a golden era for its people, an era that was to last until about A.D. 500, corresponding roughly to the age of Greek and Roman culture in Europe. It was golden for the color of the autumn fields that began to replace the woods. By a peculiar misnomer, this time of increasing farming in our northern land is referred to as the Early Woodland (500–100 B.C.) and Middle Woodland (100 B.C.–A.D. 500) periods.

Ohio farmers of this Greek and Roman time are known to archeologists as the Adena and the Hopewell people. Those names have a certain musical quality to them, but I nevertheless find it hard to call an ancient people by the modern deed to the land where their bones were first found. Whoever they were, the Adena people were the first to make pottery in Ohio, and the Hopewells the first to grow corn.

Adena (meaning "delightful place") was the name that Thomas Worthington, one of the founders of Ohio, gave to his home near Chillicothe in the early nineteenth century. A mysterious mound, 26 feet high and 445 feet in circumference, with trees growing over its slopes and top, stood silently on this land near the Scioto River. In 1901, the mound was excavated by archeologists who found, in addition to thirty-three skeletons, a great many precious tokens of devotion buried with the bodies. Among the objects was an earthen jar—the first Ohio pottery piece—copper rings and bracelets, a little raccoon fashioned of shell, and a clay pipe molded in the form of a human. The first and still foremost Hopewell site in Ohio is a burial ground containing thirty-eight mounds excavated on the land of a farmer by that name.

One clear October day, I walked across Swift's Hollow along the Vermilion River and scrambled up a nearly perpendicular ridge to a narrow point overlooking the stream. I had seen this place some years earlier from the floodplain on the other side, and remembered that its steep shale walls were sometimes referred to as the "Indian fort." Some thought that it had been a redoubt for the Eries in their seventeenth-century battles with the Iroquois, but no relics of warfare have been found there. The site, known as the Leimbach Site for the family who owns the land, dates from about 520 B.C., and hence is contemporary with other early Adena sites. It consists of a small flat terrace 75 feet above the river, but 50 feet below the general level of the ground above, hidden in its situation from all directions and approachable only by a trail down a very narrow and concealed incline from above.

One cannot go to this secluded ground without feeling that it is a special place. It seems to have been so regarded for a very long time. Although ringed with woods, the terrace itself has been kept open since the land was settled in the early nineteenth century by New Englanders, who recorded that the ground was already open when they found it. Archeologists have found no evidence that trees have ever disturbed the soil since Adena times.

We are accustomed to thinking of the American land as wilderness until the Europeans arrived. But in Greco-Roman times, farming was being carried out along the Vermilion River in northern Ohio. While farmers cultivated the remote Leimbach Site in America, Romano-British farmers developed a prosperous villa in a secluded valley of the Cotswolds in England. Like the Leimbach Site, Chedworth occupied hauntingly beautiful ground. But the Roman farm fell into disuse during the Middle Ages, and by the time that archeologists found it in the nineteenth century, it was overgrown with trees. The woods had reclaimed the Chedworth Roman villa. But in our own "wild land" of America, the Leimbach Site remained continually in use. In America as in England, humanity has been a force with nature in shaping the land through many centuries of time.

Whoever occupied the Leimbach Site for all those years was far tidier than most of us are today, for only a few projectile points and scrapers have been found. But refuse buried in shallow pits contained a few sherds of pottery and pieces of squash rind that somehow survived all these years. The humble scraps of refuse show that farming had come to northern Ohio in Adena times, and with it the manufacture of pottery. Other Adena sites have much finer pottery than ours, including large pots made for storing food in the ground, but no other place has such early evidence of growing squash. The Leimbach refuse also shows the use of many other native plants, but it is uncertain to what degree they were being cultivated: raspberries, strawberries, grapes, goosefoot, and pokeweed, as well as hazelnuts, acorns, hickory nuts, walnuts, and butternuts from the trees. Sunflower seeds have been found at other Adena sites, but no Adena site has yet shown the presence of maize.

The Adena people were the first in Ohio to build burial mounds instead of relying on beach ridges and kames. Burial sites sometimes contained precious materials from far away, such as copper from Lake Superior, indicating that trade extended over long distances. Ohio flint, from the Flint Ridge in midstate, and Ohio pipestone, a soft stone that could be carved into smoking pipes, have been found near Chesapeake Bay and in New England. With the Hopewell people, trade became even more extensive, and Ohio sites yield silver from Ontario, marine shells from the Gulf of Mexico and the Atlantic, mica from North Carolina, and obsidian and grizzly-bear teeth from the Rockies.

It is a universal human characteristic to want to leave an impression on the land, as though to show nature that we can have an effect that will remain when we are gone. The marks on the land were small until agriculture was invented; then they became much bigger. When agriculture reached northern Europe, stone circles began to be erected across the country-side, as at Avebury and Stonehenge. Silbury Hill, a huge terracing of stones covered with earth, was built near Avebury, for no apparent material purpose.

In North America, the development of agriculture was accompanied by the building of great mounds, from Florida to Canada, and from the Atlantic to the Mississippi and Missouri rivers. The greatest of these mounds were built by the Hopewell people (also called the Mound Builders), whose center was in southern Ohio, but whose culture extended from New York to Missouri, and from the Great Lakes to the Gulf.

Some of the mounds were burial mounds; others were symbolic shapes or effigies of animals, with nothing inside (as was

Silbury Hill in England). Some mounds were platforms for temples built on top; others became platforms for mounds built on mounds. The amount of earth moving required was sometimes enormous, as the mounds might cover several acres and be as much as 100 feet high. Clamshell hoes and shovels made from the scapulas of elk or other large animals were among the principal tools. (Across the Atlantic, Neolithic farmers had also used such tools at Avebury.) The greatest effigy mound in Ohio is the Great Serpent Mound, which contains no bones or relics. Much later in Wisconsin, an Effigy Mound culture developed (A.D. 700–1300), which built burial mounds shaped like animals: turtle, goose, bear, eagle, deer. Most of the mounds in Ohio are along the streams flowing south into the Ohio River.

The Hopewell people created artistic pottery, beautifully carved effigy pipes shaped like birds and toads, and figurines of clay and even fossil ivory obtained from the Ice Age mammoths. Behind this art, and the great communal building of the mounds, was maize. American corn had finally made the long genetic journey from Mexico and Central America to the shores of the Great Lakes.

At the Leimbach Site on the Vermilion River, three kernels of corn from a refuse pit, with charcoal dating from A.D. 575, show that maize was grown in our area toward the end of the Hopewell era. It is odd that so little evidence was left behind by people occupying a site for 1,100 years, for it is usual for more material to be brought into a habitat than is taken away (modern London lies many feet of refuse above the level it had in Roman days); perhaps most refuse at Leimbach went into the river below. It is fortunate that just enough was left behind to record the farming of squash in the sixth century B.C. and of maize in the sixth century A.D.

70

The Hopewell culture disappeared soon after A.D. 500, and Ohio entered its own Middle Ages (called by archeologists the Late Woodland period), which lasted until contact was made with Europeans. With the decline of Hopewell art and trade, culture in the eastern woodlands went into a dark age for 200 years. Perhaps the climate cooled enough that the new varieties of maize that had sustained the Hopewell people became less productive in the northern states. Or perhaps the decline of urban centers in Mexico affected the far-reaching Hopewell economy.

About A.D. 700, a "Mississippian culture" began to develop in the valley of the Mississippi River from Louisiana to Iowa, with large towns surrounded by palisades and supported by an agriculture based on new varieties of maize, beans, and squash. If we tend to think of the Native North American as a solitary hunter slipping through the primeval wilderness, we should be reminded of the city of Cahokia in southern Illinois.

Cahokia in A.D. 1200 had a population of perhaps 40,000 people, about the size of Philadelphia more than 500 years later, when the Constitution was signed. An enormous mound of earth was built at its center, with a base larger than that of the Great Pyramid in Egypt, and rising 100 feet high. The Americans never developed wheeled vehicles; all that soil (22 million cubic feet of it) was carried with baskets by a very large number of people working together. Their work was not finished until after A.D. 1100, at the time the Normans were throwing up mottes and building castles all over England and Wales.

The Mississippian culture probably influenced the development of the Fort Ancient culture in southern Ohio and the Whittlesey culture in northern Ohio. The Whittlesey people were farmers and fishermen, and their villages of longhouses, similar to those of the future Iroquois in New York, became increasingly year-round and permanent.

On a bluff above French Creek, close to its junction with the Black River and about 2 miles upstream from Lake Erie, is the Eiden Site, the most significant Whittlesey site in our area. It dates from about A.D. 1490, just before Columbus sailed into the Caribbean. The Eiden Site was both a village and a nearby burial ground, with remains of at least 235 burials. Most Native sites in our area are close to the lake or rivers, and because Lake Erie was one of the most productive freshwater fisheries in the world, fishing must have provided much of the food. The Whittlesey people probably became increasingly allied, in economy and culture, with other fishing people of the Great Lakes area, and less so with populations in the Mississippi watershed to the south.

The geographic position of northern Ohio, at the far edge of the watersheds of the Mississippi and St. Lawrence rivers, became of great significance to its history when European exploration of North America began in the sixteenth and early seventeenth centuries. The isolation of northern Ohio prevented or delayed direct contact with the Spanish explorations from the south, with the Dutch and English colonizations on the East Coast, and even with the French enterprises moving up the St. Lawrence River. The lack of direct contact meant that our people do not figure significantly in written documents of this period. By the time that greater contact was made with our region by La Salle and other French explorers in the late seventeenth century, the northern Ohio people had disappeared—the victims of European diseases transmitted from other Native Americans and of war with the Iroquois brought on by competition for trade relations with the Europeans. From a few written accounts and archeological findings, the last years of the "immortal Eries" (as Raymond Vietzen calls them) can be partially reconstructed, but the late Native history of northern Ohio remains largely unknown.

European penetration into the midlands of North America could be made via either of the two great river systems: the St. Lawrence or the Mississippi. The mouth of the St. Lawrence faces Europe with a large gulf that funnels sailors into the river, while the mouth of the Mississippi is remote from Europe and splayed into multiple outlets attracting little attention from the sea. As a result, most of the interior of North America was discovered via the St. Lawrence route. The Mississippi was explored from the top down, by Marquette and Joliet and then by La Salle, coming into the Mississippi watershed from the Great Lakes. When La Salle located the mouth of the Mississippi in 1682, he was able to determine its latitude but not its longitude; two years later, approaching from the Gulf of Mexico, he could not find the river that he had discovered.

The French exploration of the Great Lakes was deflected away from Lake Erie by the Iroquois of upstate New York, who had quarreled with Champlain in the early seventeenth century and who controlled the access to Erie at the Niagara portage. The French preferred to reach the upper Great Lakes by going up the Ottawa River, through Lake Nipissing, and down the French River into Georgian Bay. The people of northern Ohio inhabited a backwater away from the European traffic. Nevertheless, the Ohio people, like millions of others throughout North and South America, were devastated, in the sixteenth or early seventeenth century, by measles and smallpox and other European diseases to which they had little immunity.

For all their cultural diversity in 1492, the peoples of North and South America were biologically quite homogeneous. In blood type, for example, they were almost all type O, with some

A in the northern part of North America, and with almost no type B. They probably were all descended from a relatively small population of Asian people who had moved into isolation in the Americas before the domestication of cattle, pigs, sheep, and other animals in Eurasia. The domestication of animals in Eurasia carried a price, for a number of serious infectious agents, such as the measles and smallpox viruses, probably had their origins in cattle, pigs, and other domesticated animals and spread to humans with the close contact of domestication. The diseases of domestication became endemic in Asia and Europe, and the people of the Old World developed considerable immunity to them. The people in the New World, however, had developed no immunity to the new illnesses. They were like the large Ice Age mammals of North America: defenseless against an organism that they had never before encountered. As many of the large mammals of North America went extinct with the introduction of Eurasian hunters, so the descendants of those hunters nearly died out with the introduction of the new Eurasian diseases.

In 1519, smallpox spread from the Spanish to the American people and raged as an epidemic in the Caribbean, Mexico, Central America, and probably Peru. Nearly 100 percent of the people died in some areas. Measles, typhus, and influenza soon followed.

We do not know when the people of the southern states were first affected. From 1539 to 1542, the de Soto expedition rode through Florida, Georgia, and the Carolinas, and through the Great Smoky Mountains to Tennessee, Alabama, Mississippi, Arkansas, and Louisiana, plundering and murdering as they went. The expedition recorded dense populations of Americans, with many villages, vast agricultural fields, numerous temples on earth mounds, and centralized social authority.

The picture recorded by the English and French in the eighteenth century was entirely different, with scattered, decentralized tribes, no mound building, and much less artistic work. The entire ecology of the Southeast must have been affected by the decrease in human population. Archeological evidence is lacking for the presence of bison in the southeastern states before the coming of the Spanish, and de Soto's men saw none on their travels, but small herds were seen later by the French and English, even on the coasts of the Gulf and the Atlantic.

The epidemics in the South may have spread from one people to another up the Ohio River or across Kentucky to the people of Ohio. Or disease may have spread to Ohio from the French coming up the St. Lawrence, starting with Jacques Cartier in 1534. By 1600, the population in northern Ohio seems to have been greatly reduced, as was the population of the Iroquois in upstate New York.

For New England, historical records present a more detailed picture of the effects of European diseases on the Native Americans and their land. On the northeast coast of North America, the Native people of Nova Scotia and Maine were severely affected by several epidemics before 1616, but the denser populations along the coast of southern New England were not hit until 1616 to 1619, just before the arrival of the *Mayflower* in 1620. Squanto was the only survivor from his village. The Pilgrims found numerous fields abandoned by the Native people.

Plymouth had been an American village before it became an English colony, and more than fifty other English settlements in southern New England occupied sites that had been inhabited by Native Americans who had died of the European diseases. The English settlers were not woodsmen, and they certainly would have perished if they had had to clear a wilderness of trees to plant crops and build their homes. In fact, they found an agricultural landscape, where the woods had an open, park-like appearance because the Americans periodically cleared the undergrowth by burning. Produce from the American farms kept the English alive during the first seasons of their settlement.

The open, managed woods and fields that the English found along the coast seem to have extended across the whole of southern New England. Northern Vermont, New Hampshire, and Maine were too harsh in climate and soils to have much agriculture. The Native populations in those areas were less dense, living primarily by hunting and gathering wild foods, and the forests were continuous and thick. Northern New England corresponded to our images of the American wilderness, but southern New England was a land of villages, fields, and open woods, where it was possible to travel across the land with little more difficulty than crossing the English countryside.

When surveyors from Connecticut came to northern Ohio at the end of the eighteenth century, and settlers a few years later, they found a wilderness of almost unbroken, heavy forest. They cut trees by the hundreds and thousands to let the sunlight in for crops. We have always assumed that what they described in 1800 was the landscape in its natural condition, the way it was when the Native people lived here. It now seems probable that the Ohio land in 1600 was more like that described by the English settlers in southern New England. By the time the Connecticut people saw their Western Reserve in 1800, there had been 150 to 200 years for its fields to revert to forests and for woods previously kept open by burning to become the almost impenetrable thicket described by the surveying parties.

The population of northern Ohio was destroyed not only by

European disease, but also by the political and economic–ecological disruptions caused by the activity of European fur traders and settlers. For Americans still in the Neolithic Age, the lure of Iron Age goods from Europe was too much. Imagine what it would mean to be offered an iron kettle to cook in, if one had only pots of clay or soapstone or green birch bark, or to have an iron needle to replace those made laboriously from bone. With the coming of European goods, every Native husband would want to provide his family not only with fish or meat to supplement the crops his wife was growing in the fields, but also with beaver pelts to buy a kettle for cooking.

The Europeans wanted the fine inner fur of the beaver to make felt for fashionable hats; such are the emotive needs that drive commerce and history. The beaver had been a minor source of food and fur in the varied and self-sufficient economy of the American farmer and hunter; now it became the prime and increasingly scarce commodity in an international business competition. The Native Americans were drawn out of their previous ways of life into dependence on the beaver trade, and they began to fight with one another for access to it.

As beaver vanished from the East, eastern tribes pushed westward. The Iroquois of New York were reduced by disease to half their previous numbers, but by 1640 they had nevertheless exhausted their supply of beaver and began to make war with the Hurons in Canada and the Eries of northern Ohio. According to an account that was handed down through generations of surviving Seneca Iroquois and was revealed to a Buffalo newspaper in 1845, the Eries in 1655 decided secretly to attack the Senecas, but word of the impending attack reached the Iroquois from a Seneca woman living among the Eries. The Iroquois ambushed and routed the invading Eries at

Honeoye in upstate New York and pursued them into Ohio during the next five months.

No one knows what happened ultimately to the Eries who survived the Iroquois attacks. Some took refuge for a while on Kelleys Island, and some were probably assimilated into the Seneca nation. From 1655 until the Connecticut settlement, our region became a no-man's land. No doubt the Senecas made some use of their new territory in pursuit of the beaver, but their own population was so reduced by disease that they had no need otherwise for more land. The Iroquois nevertheless claimed dominion over the Erie land as far west as the Cuyahoga River, and sometimes roamed much farther west than that. In the latter part of the eighteenth century, the entire area from Cleveland to Sandusky may have had only 1,000 people, most of them Wyandots living in communities near Sandusky.

Above the Vermilion River, about 3 miles downstream from its junction with Chance Creek, are the remains of an extensive Erie village and burial grounds, known today as the Franks Site. The site was carefully excavated in the early 1940s by Raymond Vietzen, aided by Oberlin students and Herbert May of the Graduate School of Theology. What they found may well bear witness to the tragic last years of the Erie people. Of 187 skeletons removed from the site, about 20 percent were those of children, and 75 percent were those of people younger than thirty years. A few skeletons gave evidence of a violent death by warfare or execution, but most showed nothing but a "natural" death. The age distribution, however, speaks strongly of something unusual in nature, and it seems likely that these people were coping with the onslaught of measles, smallpox, and other European diseases. Some of the infant children were buried, heart-breakingly, under the floor of the parents' lodge (whose boundaries could be identified by the remains of post holes); in some cases, strings of beads had been placed in the children's hands. The earth is hallowed ground; the morning dew, its tears.

For hundreds of years before their demise, northern Ohio people had farmed the land for much of their food. If we were able to see Swift's Hollow in 1500, we would not see fields devoted to individual crops of corn or soybeans, but integrated fields combining corn, native beans, and squash. The fields were not plowed, but were carefully cultivated with hoes. The corn was planted in individual hills, and beans, whose vines could climb the cornstalks as though they were beanpoles, were planted around the corn. Squash and pumpkins were planted to fill out the areas between the mounds of corn and to help keep down the weeds.

There was a double wisdom in growing beans and corn together. Corn produces a large crop per acre, but requires a lot of nitrogen to do so and rapidly depletes nitrogen in the soil. Beans, with their root nodules of nitrogen-fixing bacteria, help maintain the fertility of the soil by providing the nitrogen that the corn requires. The Native farmers did not have to fertilize their fields with fish; that was an improvisation required by the English when their corn did not grow in beanless fields.

The beans also aided the nutrition of the people who ate the corn, when the two foods were combined as succotash. The protein of maize is deficient in tryptophane and lysine, two of the eight essential amino acids that must be present in the diet if our bodies are to make the enzymes and other proteins we require for life. Tryptophane can also be used by our bodies to make nicotinamide, a vitamin scarcely present in corn but essential to many biochemical reactions. European peoples who became too dependent on corn alone in the eighteenth and nineteenth centuries became sick and died of pellagra, due to a lack of nicotinamide. The Native Americans knew better than to rely on only one food, and the beans contained the tryptophane, lysine, and nicotinamide that were lacking in the maize.

The Native people of the Great Lakes area were far more intimately acquainted with their local flora than most of us are today. Between 1907 and 1925, Frances Densmore, a graduate of the Oberlin Conservatory of Music, traveled to Minnesota, Ontario, and Wisconsin to study the music of the surviving Chippewa (Ojibwa) people there, but while she was at it, she also recorded the uses they made of wild plants. She cataloged nearly 200 plants used by these people alone. More recently, R. A. Yarnell has recorded 130 plant species used for food by the people of the Great Lakes area, 275 used for medicine, 27 for smoking, 18 for beverages and flavorings, 25 for dyes, 31 for magical charms, and 52 for other purposes. One extraordinary application was reported by Peter Kalm, a Swedish botanist who visited America in 1748 to 1751. According to Kalm, the Seneca people soaked seeds in a poisonous extract of the American false hellebore, before throwing them into their fields for the birds; the disoriented behavior of the birds that ate the seeds discouraged other flocks from visiting the field and eating the crops.

The North Americans were able to domesticate only the dog, among mammals, and so for meat they were dependent on hunting. The black bear, which must have been numerous then, was a favorite food, as was the whitetail deer. Bear and deer can be lured close to human habitations and so become easy prey for anyone with a bow and arrow. Indeed, bears are frequently uninvited guests wherever there is food or refuse. I have seen enough of their tricks at hiking shelters in the Smoky Mountains, and at tent sites in the Boundary Waters, to understand how they could become a favorite target. Mountain lion, wolf, elk, and bison were also hunted in Ohio, and among the birds, passenger pigeon, turkey, grouse, quail, and Canada goose. Hawks were trained to keep birds out of the cornfields, and turkey and ducks may have been domesticated.

Across its range in North America today, the raccoon has a special relationship with people. A friend had one in his chimney the other day. Last winter, one perished in an unseen corner by my front porch, seeking shelter from a blizzard. My compost pile sees a traffic in raccoons every night, and sometimes some noisy squabbles. The storm sewers of Oberlin are raccoon dens, and after a heavy rain, a forlorn masked face may be at a curbside drain. I suspect that the tameness of the raccoon today is testimony to a long history of close interaction with the Native people of America, a closeness that may have approached domestication.

In the seventeenth century, the Eries of Ohio were known, according to Jesuit missionaries living across the lake with the Hurons, as the "Cat People," because they wore the skins of "cats" for clothing. At the Franks Site on the Vermilion River, Raymond Vietzen found large numbers of raccoon bones, especially the penis bone, which must have been viewed as possessing fertility or other power. Vietzen believes that the Eries made clothes from raccoon skins and that the Jesuits did not distinguish the raccoon from a cat. The Eries were probably the "Raccoon People," and may even have raised raccoons for their skins and meat, as farms raise mink for their skins today.

For hunting, seasonal migration, and trade, the native Ohioans had both water routes and trails. The Cuyahoga River was used for reaching, via a portage, the headwaters of the Tuscarawas River and the way south to the Ohio River; this was probably the most commonly used route in our area for connecting the Great Lakes with the Ohio and Mississippi rivers.

Other routes were along the Black River to the Killbuck River and via the Vermilion River to the Jerome Fork of the Mohican River. All these routes come together near Coshocton as the Tuscarawas, Killbuck, and Mohican (with the Walhonding) form the Muskingum River. Important overland trails included a Lake Trail along the Lake Erie bluffs and a Watershed Trail along the high ground from Kent through Medina, passing a few miles south of Wellington and going on to New London and Milan. There was also a Mohican Trail along the Vermilion River through Birmingham and Wakeman, heading for the Mohican River. The glacial beach ridges provided local trails.

When Columbus first made contact with the people of the New World in 1492, he had no idea who they were or where he was. He had come from a land of terrible warfare between Spaniard and Moor, where religion annointed the plunderer and oppressor, where barbaric cruelty to animals was extended without hesitation to people, and where widespread famine, disease, and hardship gave rise to a violent and malevolent view of nature. It was not much different in England or France or anywhere in Europe in the fifteenth century. For a few brief weeks in October and November 1492, there was a chance that Europe would see the true riches of a new land in the spirit of its people. Columbus was happily surprised by the beauty, the strength, and the kindly, generous, and loving character of those who met him and helped him in every way they could. There was just a slight chance for the Old World's salvation, but the Europeans, alas, were not to be converted. Their ships were laden with firearms and their souls with greed. They wanted to possess whatever they could find.

In the county administration building in Elyria, Ohio, a deed is recorded for my home in Oberlin. It describes, in trigonometric surveying terms and in units invented across the sea, a tiny piece of the earth's surface, and recounts the transfers of this parcel from one party to another. Across all the Americas, in every county or administrative region, are similar monuments to the recent "ownership" of the New World. Yet our homes are but tents on the landscape of time, and we but visitors to a world whose age exceeds our own 100 million times. We own only what the spirit creates. As Tecumseh well knew, we cannot possess the land or the sea or the sky, but we can treasure an awareness of nature's long history and hold fast the reverence of hundreds of human generations before us.

4 western reserve

On a gravel terrace above the Thames, a few miles down-stream from London, stands the old Greenwich Observatory, with its sighting telescope, or transit, mounted on a north–south line that the world has come to accept as the prime meridian of the earth, the line of 0 degrees longitude. The Greenwich Observatory was built in 1675 and 1676 because British ships were sailing all over the globe without knowing exactly where they were, or which direction was home, and because British colonies were being settled in the New World without knowing where they were either.

Our earth has no coordinates scribed on its surface. Land features can identify places, such as the hill at Greenwich and the beach ridges north of Oberlin, but there is no geometric guide to distances or directions between these features. In the heavens, however, there is a map, defined by the patterns of the stars. Those patterns project on the earth as though from one spherical surface high above the other. Wherever a person stands on the surface of the earth, there is one point in the celestial patterns that is directly overhead, and that point is different for each location. By taking some particular point as an origin, we can construct, by measuring angles away from that origin, a grid of latitude and longitude in the heavens, which then can be projected on the earth. Each point on the earth has a unique latitude and longitude; nobody else's house has the same latitude and longitude as yours. But to find out what it is, we have to look to the map in the heavens.

The trouble with this celestial map is that it moves. It moves because the earth moves, with daily rotation about its axis and with yearly revolution around the sun. The movement of the earth around the sun is not troublesome, because it affects only the position of the sun against the pattern of the stars. But the rotation of the earth changes the east–west position of the whole pattern relative to any point on the earth. If we move from one point on the earth to another, we do not know how much of the east–west change in the pattern is due to our change in position, and how much is due to the rotation of the earth while we changed position—unless we have an accurate clock with which to keep track of how much time the earth has had to rotate while we traveled from one place to another. Thus the longitude of a ship at sea or of the colony at Plymouth, Massachusetts, relative to that of London, could not be determined without a clock that could be set at London and counted on to keep accurate time for the duration of the voyage at sea or the duration of the trip to Massachusetts.

Latitude, or north–south position, was not affected by the turning of the earth, and could be read from the elevation of the pole star above the horizon or of the sun at noon on a specified day. In the seventeenth century, ships and colonies could know their latitude but not their longitude. Ships finding their way home would sometimes get on the latitude of their home port, and then sail due east or west until they reached home or ran aground on something in between. Columbus in 1492 sailed due west from the Canary Islands, keeping a course at about 28 degrees north latitude across the Atlantic. He could keep his latitude constant by measuring the angle of the North Star above the horizon, but it was initially also helpful that he left the Canaries on the autumnal equinox, when the sun would rise and set due east and west. Columbus could determine his north–south position on the earth, but he could not know how far he had traveled west without an accurate clock.

The moon, however, is a kind of clock in the sky, for its motions are independent of the earth's turning. The Greenwich Observatory was built to track the motions of the moon so accurately that predictive tables could be published that could be carried from one place on earth to another to be used as a clock. By watching the position of the moon against the stars, one could tell how much time had elapsed since leaving London. One could then subtract the effect of the earth's rotation from an observed change in the east–west position of the stars to determine one's longitude. By such a clever device, the building of an observatory in Greenwich could tell the people in Plymouth where they were—relative to the prime meridian in Greenwich. (Of course, the Native Americans who lived in Plymouth before the English arrived knew exactly where Plymouth was; they just did not know where Greenwich was.)

The Greenwich Observatory had not yet been built when the royal charters were granted to the early English colonies on the Atlantic seaboard. No reference to longitude could therefore be made in defining the boundaries of the colonies. In the charter for Virginia granted in 1609 by James I, no reference was made to latitude either. The colony was to have 400 miles along the Atlantic coast, centered at Old Point Comfort, with its southern boundary going indefinitely due west, and its northern boundary slanting northwest all the way to the Arctic. The future Ohio was fully within the boundaries of Virginia. But in 1620, part of it was placed in Massachusetts as well, as James I granted to Plymouth all land between latitudes 40 and 48 from the Atlantic to the Pacific. In 1664, Charles II gave to his brother, the Duke of York (later James II), land extending from Delaware Bay to the St. Croix River, which gave New York a claim to Ohio also. The English had no idea of the immensity of the land that they were trying to claim and to parcel up. Sir Francis Drake had reported that he had seen both oceans at once from the mountains of Panama, which may have encouraged the belief that there was no great distance anywhere between the Atlantic and the Pacific. Even as late as 1740, the Duke of Newcastle addressed a letter to the "Island of New England."

The future of the Ohio land was to be greatly influenced by such colossal ignorance of geography. Aliens who knew nothing about the American land nevertheless claimed to own it and to divide it among favored parties, without taking into consideration those whose ancestors had lived on the land for 11,500 years. In 1630, a year after he had dissolved the English Parliament, Charles I conveyed to the Earl of Warwick and his Council of Plymouth the land that was to become Connecticut and its Western Reserve. The language of this grant seems calculated to create an image of precision from utter confusion, by a repetition of undefined but high-sounding phrases. The colony is described as

> all that part of New England, in America, which lies and extends itself from a river there called Narragansett river, the space of forty leagues on a straight line near the sea shore,

towards the south-west, west by south, or west, as the coast lieth, towards Virginia, accounting three English miles to the league, all and singular, the lands and hereditaments whatsoever, lying and being within the bounds aforesaid, north and south in latitude and breadth, and in length and longitude, and within all the breadth aforesaid throughout all the main lands there, from the western ocean to the South Seas.

The charter was amended by a further one in 1662 from Charles II to the governor and company of the English colony of Connecticut. The eventual interpretation was that Connecticut extended for 120 miles along the coast from Narragansett Bay to a point on the forty-first parallel, then north to 2 minutes past the forty-second parallel, and then due east again. But the colony also claimed lands westward "throughout all the main lands there." Connecticut people settled in New York and Pennsylvania, and disputes arose over jurisdiction.

The Continental Congress asked the colonies to give up their western claims. New York did so in 1780 before the Revolution was concluded, and under the Articles of Confederation, Massachusetts gave up its claims in 1785. In 1784, Virginia ceded governance over everything northwest of the Ohio River, but retained ownership of lands between the Scioto and Little Miami rivers, to be able to pay in land Virginia veterans of the Revolutionary War. Connecticut was the last to give in, and it did not give in completely. The state surrendered in 1786 only those lands lying more than 120 miles west of the western border of Pennsylvania, reserving a western tract of land that was a kind of mirror image of its eastern self. This land, which became known as the Western Reserve of Connecticut, was to have Lake Erie as its northern boundary, as

Connecticut had Long Island Sound as its southern boundary. It was to lie, like Connecticut, between 41 degrees on the south and 42 degrees, 2 minutes, on the north. And it was to extend for a distance of 120 miles, just as in the 1630 charter for Connecticut, except that the distance was to be measured not along the shoreline, as in Connecticut, but due east and west, starting at the Pennsylvania border.

By the 1780s, longitudes (unknowable in 1630) could be measured with the aid of accurate clocks, and state lines could now follow meridians of longitude as well as parallels of latitude. In the case of Pennsylvania, it was decided that the state's western boundary should follow a meridian exactly 5 degrees west of where the Delaware River on the east flowed across the southern boundary surveyed by Mason and Dixon. That worked out to give Pennsylvania a western boundary, and Ohio an eastern one, along the meridian lying 80 degrees, 31 minutes, west of Greenwich. The Delaware River between Pennsylvania and New Jersey thus determined where the eastern boundary of the Connecticut Western Reserve would be. But because the meridians converge toward the pole, the western boundary of the reserve, defined as always 120 miles west of the eastern boundary, had to deviate about 4 degrees west of true north.

Insofar as Connecticut claimed sovereignty as well as ownership of its Western Reserve, it stood in defiance of the Northwest Ordinance adopted by Congress in 1787, which set up the Northwest Territory under federal jurisdiction. In fact, Connecticut never attempted to govern the reserve, which was much too distant to be under its control. In 1788, the state sold about 25,000 acres near Youngstown to General Samuel Parsons of Middletown, Connecticut; the tract was considered valuable for the salt spring that the Native Americans used, but the spring proved to be of poor quality, and Parsons died before he could develop it.

In 1792, Connecticut designated the most distant part of the reserve, the westernmost 25 miles, as land to be given to Connecticut citizens whose coastal homes had been burned by the British (led in part by a Connecticut native, Benedict Arnold) during the Revolutionary War. This land became known as the Firelands. The remainder of the reserve, 95 miles from east to west, was sold in 1795 for $1.2 million to fifty-seven private individuals who formed an unincorporated association known as the Connecticut Land Company. The funds from the sale went into an endowment for public education in the state. Connecticut ceded all sovereignty over the reserve to the federal government in 1800. The Western Reserve became part of the new state of Ohio in 1803.

Who owns the land on which we live? In 1795, fifty-seven persons in New England claimed to own nearly 3 million acres of a land they had never seen. They said that it had been sold to them by Connecticut. Connecticut's claim went back to the ignorance and arrogance of English kings who said that the New World, whatever it was, was theirs because John Cabot had "discovered" it in 1497. Those who already lived there somehow did not count, because they did not live like Englishmen and did not understand what it could mean to sell the land. The Natives viewed their American land much as the ancient Celts had viewed the land of Britain: that it was owned by all who used it. Across the English land today are thousands of public footpaths testifying to the time when property rights were understood in Britain much as the Natives regarded them in America. It has been our great loss that American common rights were destroyed by traditions of ownership descended from the feudal views of Norman conquerors, who thought that a king could own a country.

The few surviving descendants of the Native Americans of northern Ohio were defeated by General Anthony Wayne at the Battle of Fallen Timbers on the Maumee River in 1794. In the resulting Treaty of Greenville of 1795, they gave up their lands east of the Cuyahoga River. Ten years later, on July 4, 1805, tearful Natives accepted $4,000 at Fort Industry, with promises for $14,916.67 more, in exchange for all their lands in the Firelands and the reserve west of the Cuyahoga River. The unhappiness generated by the Treaties of Greenville and Fort Industry helped convince Tecumseh and his brother that a Native confederacy was needed to resist the encroachments of the whites, but William Henry Harrison dispersed them at Tippecanoe in 1811, and Tecumseh, allied with the British, died two years later in the War of 1812. Tecumseh believed that it made no more sense to sell the land than to sell the ocean or the sky.

To divide the land among its members, the Connecticut Land Company in 1796 sent one of its directors, Moses Cleaveland, to survey the reserve from the Pennsylvania line to the Cuyahoga River. His men were unable to complete the survey that summer, and so returned in the summer of 1797. Prior to the Treaty of Fort Industry, the Native Americans regarded the Cuyahoga as the boundary of the United States, and so the survey of the land west of the Cuyahoga was not done until 1806 and 1807.

In their summers of laboring amid clouds of mosquitoes, slogging across streams and through swamp woods, the surveyors shaped forever the way future generations would live on this part of the earth. The science of latitude and longitude was to be applied to the earth. Instead of living on the land and working with its natural features to decide the boundaries of estates, based on common and individual needs, the land was to be measured as so many squares to be apportioned among the partners of the company. Perhaps because the Firelands had already been allotted an east–west extent of 25 miles, it was decided to measure the land into squares 5 miles on a side, instead of 6 miles, as was being used elsewhere in the state. All townships and counties in the Western Reserve and Firelands, and many of the roads, would be different today if the decision had been for a 6-mile rather than a 5-mile grid.

The Cleaveland surveying team started its work by finding the western boundary of Pennsylvania at Lake Erie. The Pennsylvania boundary had been surveyed and cleared of trees in 1786; ten years later, it was overgrown with young trees, but Cleaveland's men could still follow it southward to the forty-first parallel. They erred southward about three-quarters of a mile in the company's favor, for their starting point was actually at latitude 40 degrees, 59 minutes, 21.6 seconds as they began to survey the southern boundary of the reserve.

The surveying plan was to delineate twenty-four north–south range lines at 5-mile intervals west of the Pennsylvania line. The columns of land between these range lines were called ranges, Range 1 being the easternmost of the Western Reserve, and Range 24 the westernmost of the Firelands. The ranges would be cut at 5-mile intervals by township lines running east–west, and the squares thus formed (of 25 square miles) would become individual townships. By a confusing convention, the rows of townships are also called townships, so that Township 1 refers to the first row of townships north of

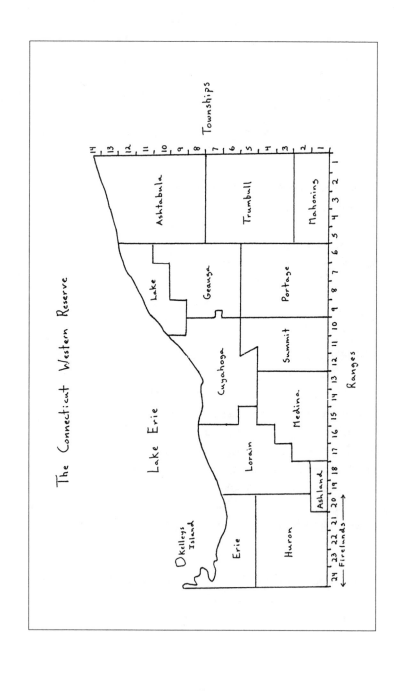

The Connecticut Western Reserve

the southern boundary of the reserve. Russia Township, for example, which contains Oberlin, is the intersection of Township 5 with Range 18. Because the Lake Erie shore runs in a southwesterly direction, the number of full townships is 13 in Range 1, but only 6 in Range 18.

The surveyors had enormous difficulty, in 1796 and 1797 in the eastern part of the reserve, and again in 1806 and 1807 in the western. They had to cut their way through thick forests to make their sightings with their compasses and their measurements with their chains. They were in a hurry to get the job done, and substantial errors were made in both sightings and measurements. They apparently attempted to make the range lines true meridians north and south, until on the western boundary of the Firelands they realized that they would be giving the entire reserve less than its full 120-mile width if they did not run the final line nearly 5 degrees west of north, to make up for the convergence of meridian lines toward the pole. The western range line for Range 19 was also adjusted west of north. There was a tendency to err on the generous side in measuring the ranges, with the result that Range 19 came out with much less than 5 miles to give the Firelands a full 25 miles for its Ranges 20 to 24. Some of the difficulties and frustrations of the surveyors can be felt in the annotations made on their 1808 map of the Firelands: "entered a g——d d——m Bad Willow Swamp," "this land not worth a farthing," and "set a post in hell." Theirs was not a script for a future chamber of commerce!

A summer swamp can be a dismal place for someone with a purpose. Having slogged through muskeg and swarms of mosquitoes, trying to find the end of a northern portage, I shudder for the men who had to align their long surveyors' chains, measuring and recording their way through the swamp woods of northern Ohio. But go to such woods in winter with no purpose but to see what the wildness of the world is up to, to measure the progress of the skunk cabbage pushing up through the half-frozen mud of early March, or to discover the bright green shoots of the *Mnium* moss springing forth atop a rotting log, and the same woods holds a different world. I went to the swamp woods one late winter day to consider the reflections in the forest pools. Black ash and red maple stood as double images amid the icy waters, drinking of the pools and of the rising mists. Half-drops of water hung from finely curved hornbeam twigs. Pools were tiled with fallen leaves and timbered with fallen trees. Rooted in an old elm log, a young leafy herb hung its green banners across the melting winter, reaching for the first light of spring. In late wintry silence, there is no greater beauty than the accursed woody swamp.

The surveyors aligned our lives with the coordinates of our planet, so that lines pointing toward the poles or running parallel to the equator are everywhere in our world. They could not have got away with it quite so successfully if the glacier had not preceded them. But on a flat landscape, it was possible to cut straight roads on or parallel to the range and township lines. Ohio Route 58 proceeds, over much of its course, straight north and south up the middle of Range 18. The villages of Sullivan, Huntington, Wellington, Pittsfield, and Oberlin are all on this north–south line. Except for Oberlin, each is at the midpoint of its respective township, each 5 miles from the next one. To the east, Route 301 runs up the middle of Range 17, with the villages of Homerville, Spencer, Penfield, and Lagrange at 5-mile intervals; and Route 83, in the middle of Range 16, similarly has Chatham, Litchfield, and Belden. It is remarkable that Ohio villages were founded at points specified not by the character of the land itself, but by a distance measured in an ancient English unit from an alien line (the Pennsylvania boundary) that was constructed 5 degrees of converging earthly longitude from where the Delaware River, in its meanderings, happened to flow across the Mason–Dixon Line.

Within the villages, the streets usually run due east and west, or north and south, and the houses also are so aligned. Like the rocks at Stonehenge, they record the hours and days of the sun. The autumnal equinox is coming, and the sun is rising straight out the east window of my kitchen. When I bicycle along College Street in the evening, the sun will be sinking directly into the end of the street. Spring and summer are when the sun rises and sets north of my street, and fall and winter, when it stays always south of the equatorial parallel paved in front of my house.

A few towns defied the geometers' grid because of significant topographic features of the land. Rivers, with their water and power, and sandy beach ridges, with their ancient roads and well-drained soils, could hardly be ignored. Elyria was founded where the east and west branches of the Black River form waterfalls and come together. Brownhelm and South Amherst and North Ridgeville were placed on the sandy lake ridge trails of the earlier Americans. No one had heard of continental glaciers in 1817 when Brownhelm and Elyria were founded, but the action of glacial ice and waters had left impressions on the land that drew people to those places.

Accompanying the surveyors in 1796 and 1797 and 1806 and 1807 was a committee of land assessors who made notes on the character and value of the various townships. The townships of 25 square miles were generally divided into 100 lots, so that an average lot had one-quarter of a square mile, or 160 acres. A scheme was devised for equalizing the value of townships by adding to the less valuable ones tracts of land from elsewhere in the reserve. To make plenty of such tracts available, a number of good-quality townships were designated as "equalizing townships," to be cut into parcels for distribution. Fortunately, of course, it was only the ownership of the land that could be distributed, and the land itself stayed happily where the glacier had left it. In Lorain County, for example, the townships of Range 19 south of Brownhelm (Rochester, Brighton, Camden, and Henrietta), all of which were considerably less than 5 miles wide because of surveying problems, were designated for equalizing purposes, as was the Black River land between Amherst and Lake Erie. The equalizing committee was not impressed with the remaining townships of Lorain County, for all of them were given equalizing parcels to bring them up to the standard value. Wellington received 4,000 acres from Brighton; Pittsfield, 3,000 acres from Camden; and Russia, 4,300 acres from a tract in Range 12. Even the islands in Erie were pulled out of the lake by the accountant's ledger: the Bass Islands were given to Avon, and Kelleys Island to Carlisle!

With the surveying and equalizing done, all those who held shares in the Connecticut Land Company awaited the drawing to see who got what. Two members of the company, Titus Street and Isaac Mills, had poor luck: They drew Russia Township and its equalizing tract from Range 12. Mills later sold his interest to Samuel Hughes. So two men in Connecticut who had never seen Ohio now "owned" about 28 square miles of it—a land that had taken 500 million years to make, the domain of ancient fish and mastodons, and the home of 400 generations of Native Americans now vanished from their beech–maple woodland and their fields of green and golden maize.

Ohio, as a state emerging from the Northwest Territory beyond the Appalachians, was a union of the western colonial interests of Virginia in the South and Connecticut in the North. Virginia (which then included West Virginia) was just across the Ohio River, so its people got to Ohio first. In 1796, a road was cut by Ebenezer Zane from Wheeling to Zanesville and down to Maysville on the Ohio River, to facilitate travel and settlement in the southeastern part of the territory. The impetus for statehood came from the Virginia settlements. In 1803, Ohio became a "southern" state, but it was about to become a northern one.

In 1796 in the northern part of the territory, the surveyors built log cabins on the present site of Cleveland; as payment, some were granted land to the east for their own settlement, which they honored with the name of their master geometer, Euclid. So a few New Englanders began to live east of the Cuyahoga in the late 1790s, but the settlement of land west of the Cuyahoga had to wait until after the Treaty of Fort Industry of 1805 and the surveys of 1806 and 1807.

Lorain County had its first settlers in 1807, when three persons settled in Black River and six hardy souls from a party of thirty reached Columbia Township from Waterbury, Connecticut. The remainder of the Waterbury party arrived in the spring and summer of 1808. Vermilion had its first settlers in 1808, Eaton and Ridgeville in 1810, Avon in 1814, and Sheffield in 1815. But people were slow to come to northern Ohio from New England until a volcano erupted on the island of Sumbawa in the East Indies.

The landscape of New England today shows that its farms were already in trouble before Mount Tambora erupted on April 5, 1815. It is easy to distinguish an ancient woodland, with its rich diversity of native plants, from a younger and less varied woods that has grown back from open fields. Southern New England has an abundance of young woodlands, testifying that farms were once more prevalent than they are today. In the mid-nineteenth century, nearly 75 percent of Connecticut was open country; 100 years later, woodlands claimed 63 percent of the state. In southern New Hampshire, stone walls built to be the boundaries of fields now run through deep woods, covered with mosses and ferns. The population of Hartland Township in Connecticut had already begun to decline in 1790, as farmers gave up on stony soils and uncertain growing seasons.

And then it happened. Tambora exploded, killing tens of thousands of people in the East Indies (the most deadly eruption of modern times) and throwing enormous quantities of dust into the atmosphere. The dust reflected sunlight, and weather was affected throughout the world. In New England, the summer of 1816 was no summer at all. Jabez Hanford recorded in his Bible that his ground was frozen on June 6 to 10 and that there were frosts again on July 24, August 27, and August 29. The crops were negligible except along the coast, and many families nearly starved. It took little persuasion to convince them that life might be better by the moderating shores of Lake Erie, and the summer of 1817 saw a large migration from New England to the Western Reserve and Firelands of Connecticut.

Many townships and villages were first settled in the years immediately following the eruption of Mount Tambora: Grafton in 1816; Brownhelm and Elyria in 1817; Amherst, South Amherst, Russia Township, Penfield, Wellington, and Huntington in 1818; Carlisle in 1819; Brighton in 1820; and Pittsfield in 1821. The village of Oberlin had to wait until

1833, but since both Brownhelm and Elyria were staging points for the creation of Oberlin, the college village too goes back to the 1817 exodus from New England.

The settlement of Brownhelm was led by Henry Brown of Stockbridge, Massachusetts. He came to the township in 1816 with six young men, including Peter Pindar Pease, who became the first resident of Oberlin seventeen years later. The men built a log house, and Pease and three others stayed the winter to continue clearing trees, while the rest returned to Massachusetts to lead a group of families to Brownhelm the following summer. The families arrived on July 4, 1817, so that Brownhelm celebrates its anniversary on the same day as the nation. The next year another group arrived, including the family of Grandison and Nancy Fairchild, with their young children, Charles, Henry, and James (who had been born only the previous November). The parents were schoolteachers as well as farmers; Grandison had taught young Mark Hopkins, who became president of Williams College. After attending the new college at Oberlin, Henry and James would also become college presidents: Henry at Berea College in Kentucky, and James at Oberlin. On their trip to Brownhelm, the Fairchild family traveled overland from Stockbridge to Buffalo, and then by water to Cleveland on the *Walk-in-the-Water*, which had just been launched as the first steamship on Lake Erie.

The Brownhelm settlers spread themselves across the township, from the lake shore to the higher ground of Henrietta, with many families on the beach ridges of the ancient Lakes Warren, Whittlesey, and Maumee. No one had any idea that the ridges had been built by postglacial lakes; Louis Agassiz was still a boy of ten in far-off Switzerland. The main village of Brownhelm was settled along the distinctive Whittlesey ridge.

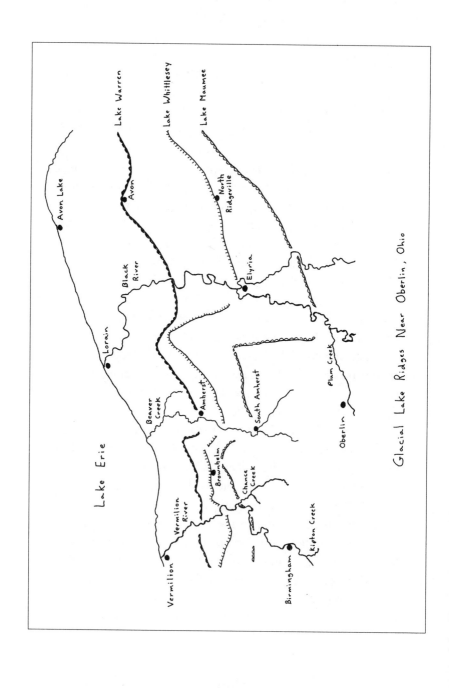

Glacial Lake Ridges Near Oberlin, Ohio

Its old houses today look out on a scene molded by millennia of human and natural history. On the shadowed ridge at sundown, I think I almost see the smoke of ancient campfires, the hunters fashioning their fluted spear points, and mastodons browsing where now there is maize.

A vivid account of pioneer life on the Western Reserve was written by James Fairchild for the fiftieth anniversary of Brownhelm. In 1817, the Brownhelm region was a thick forest, for it had been nearly 200 years since the land had been farmed by Native Americans. Fairchild describes the vast labor of felling trees to clear spaces for log houses and crops; the isolation from external commerce, especially in the years before the Erie Canal was built across the state of New York; the lack of any money to buy goods from the eastern states; the howling of wolves at night; and the construction of the first frame or brick houses in the village. The Fairchild family had to provide everything by their own hands, growing their own food while clearing the woods, and endeavoring to organize a school for the village children. Nevertheless, the Fairchilds were able to move from a log house to a brick one in just six years and to educate two future college presidents.

Fairchild's portrayal of the original Brownhelm forest is perhaps the most beautiful one we have of the native woods of the southern Great Lakes region. It evokes images of magnificent trees, like those still standing in hardwood coves of the Great Smoky Mountains:

> Along the ridges the chestnut prevailed, the trunk from two to four feet in diameter and a hundred feet in height, furnishing the best fencing material that any new country was ever blessed with. . . . The tree next in value for timber was the white-wood

or tulip tree—of regal majesty, and second only to the white
pine for finishing lumber, and for some uses superior to it. The
oak and the hickory in every variety and of magnificent pro-
portions, were found every-where; and, on the low-lands and
river bottoms, the black-walnut, probably the most stately tree
of Northern Ohio forests, inferior in magnificence only to the
famous Red-wood of California. A single specimen was standing on the Vermilion River bottom . . . which was said to measure 15 feet in diameter.

The clearing of the forests of the Western Reserve created
an unexpected hazard for farmers with domestic livestock
(the Native Americans had none). White snakeroot is native
to the Ohio woods, and it can become abundant in late summer in the sunlight at the margins of woods and fields. Grazing cattle will often eat the white snakeroot at the edge of the
field, especially in dry years when the grasses are sparse. The
plant contains a toxic substance, tremetol, which gradually
builds up in the bodies of the grazing cattle. Before the toxin
has reached levels sufficient to cause alarming spasms in the
farm animals, it is already present in their milk in concen-
trations that can be lethal for people. A disease called milk
sickness, or trembles, was first noted in North Carolina and
became prevalent in the Midwest as farmers with cattle moved
in. In the mid-nineteenth century, a poisonous plant was sus-
pected, but it was not identified until 1917. As Fairchild
recalled the problem:

> The disease most dreaded in the new country was the milk
> sickness . . . commonly supposed to originate in some poiso-
> nous herb eaten by the cattle, and to be communicated by the
> use of milk. . . . No part of the town was entirely exempt, but

the disease was developed especially in certain localities. The Barnum place, near the old meeting house, was remarkably afflicted with it, and three stones side by side in the burying ground mark the graves of three Mrs. Barnums. . . . One autumn four members of their family died within a week. . . . The latest calamity of the kind was in 1838, when the entire Campbell family, of five persons, died in the space of a month.

In southern Indiana, a few miles from the Ohio River, another wilderness family was struck by milk sickness in the autumn of 1818. Nancy Hanks Lincoln died at age thirty-four, and Abraham (nine) and his sister Sarah (twelve) had to fend for themselves for a time without a mother.

On the very day that the first families reached Brownhelm from New England, ground was broken at Rome, New York, for the construction of the Erie Canal. The work took nearly nine years to complete. The migration of New England families to the Western Reserve became much easier when the canal opened in 1826, for the entire trip from the Hudson River to Cleveland could then be made by boat.

The Erie Canal also made Ohio more attractive to farmers, and New England less so. Before the opening of the canal, the only product that could profitably be carted out of northern Ohio was potash, made from burning those beautiful trees of Brownhelm and of other new communities. Now Ohio crops could be sent east, with transportation costs across New York State reduced from $100 to $5 a ton. Ohio wheat could be ground to flour at the mills of Buffalo or Rochester, and sent on to New York City, or even to the West Indies. New England farms now had to compete with the favorable soils and climate of the southern Great Lakes.

Missionaries, as well as merchants, began to turn to the Middle West. The land of Russia Township was to feel the calls of religion in the wilderness, with results as dramatic for its swampy woods as the building of Avebury or York Minster in England. In 1831, while the young theology graduate Charles Darwin went geologizing with Adam Sedgwick in Wales, and then set sail on the *Beagle*, John Jay Shipherd brought his family from Vermont to Elyria and began to serve as a missionary pastor. A letter to his father in April 1831 described the countryside (and the people) he found:

The face of the country is plain, crossed in various directions with sand ridges. From these ridges of light soil you gradually descend to the low grounds, which are clay or heavy loam, too wet for plowing [as often they are even today in April] but fine for grass. These lands are heavily timbered with chestnut, oak, white wood, hickory, maple & Beech, ash, etc. I have not seen a pine in the country. White wood is a good substitute. . . . The four miles . . . lying north of this village are almost unbroken wilderness. A few families are scattered through it, & wolves enough. Returning from a mission among these families, night overtook me—I lost my way, as I could not see the marked trees or tracks which were covered with leaves—& to comfort me while searching for the road a gang of wolves set up a howling which make the woods ring . . . no one here has ever been injured by the wolves. Wolfish men are much more to be dreaded.

Shipherd soon became discouraged with his work among the wolfish men. He nearly quit, but then revived himself with a new plan to found a colony and college devoted to Christian teachings and totally free from sin. (An easier goal was set by Darwin in trying to understand the whole history of life on earth!) Shipherd enlisted his friend Philo P. Stewart, who had been on a mission to the Choctaw people in Mississippi, and together they searched for a site for their colony. They seemed to have wanted a place that was accessible by road but empty of people for miles around, and so, in effect, they were looking for a spot that earlier settlers had shunned. They found it on the north–south road through the middle of Russia Township where a rough track came in from Elyria, a spot where sin would have to travel at least 3 miles to reach them from any direction.

They knew that the land was for sale. They had no money, but believed that their plan was so noble that God would provide for them. Shipherd traveled to New Haven, where the owners of the land, Titus Street and Samuel Hughes, proved willing to give some of it away in exchange for getting something attractive started where nobody had previously wanted to go. Shipherd obtained a gift of 500 acres for his college, provided that within three years he could erect buildings worth $5,000 and enroll fifty students in his school. Farmers, tradespeople, and teachers attracted to the school could purchase (at $1.50 an acre) land of their own from the 5,000 acres retained by Street and Hughes.

Religion had given birth to a fine commercial deal for all, provided one could cut trees into classrooms in three years. There was no time for faculty committees! Shipherd traveled to New England and New York to find students, teachers, and money, while Stewart hired Peter Pindar Pease of Brownhelm to begin cutting the trees of Oberlin in early April 1833. Pease built a log cabin on a small tributary of Plum Creek, and on

April 19 moved his family into it; less than two months later, ten other heads of families had joined him, and 20 acres had been cleared of trees. A rectangular college campus (now Tappan Square) was laid out, ten surveyors' chains wide by thirteen long, parallel to the north–south road.

I walked across Tappan Square one morning, after an April rain, and thought about the contingencies of history beneath a grassy pool of water glistening in the square. The pool is all that is left on the surface of the little stream that ran beside Pease's cabin, 160 years ago. Beneath the square, the stream's waters are now carried by a conduit down to Plum Creek. Like the Walbrook and other lost rivers of old London, the flowing waters of our history sink into the land. Tappan Square (actually a rectangle) is 660 feet wide, which is ten chains or one furlong ("furrow long"), which was supposed to be the average length of a plowed furrow in the fields of medieval England. Each of the thirteen chains of its length gives ten square chains in area—an English acre, or the amount of land a medieval oxen team might be expected to plow in a day. This village square is aligned with the earth's equator and is edged by a meridian in the middle of the eighteenth 5-mile range west of the Pennsylvania–Ohio border.

Why is a place where it is? The medieval colleges of Oxford grew up by the walls of an old Saxon market town, which centuries before had developed on a gravel terrace barely above the confluent waters of the Thames and Cherwell rivers. Shallows of the Thames provided a ford for oxen driven across the Saxon countryside, and so the town received its name. Generations of scholars have come from afar to study, to dream, and to write down new thoughts, at the place where cattle once crossed the rivers. But a little college village in the Western

Reserve is here because the kings of England had no idea what lay beyond the Atlantic shores of the New World, because a river eroded the course it did in flowing south to Delaware Bay, because a volcano blew up in the East Indies, and because an observatory was built on the hill at Greenwich.

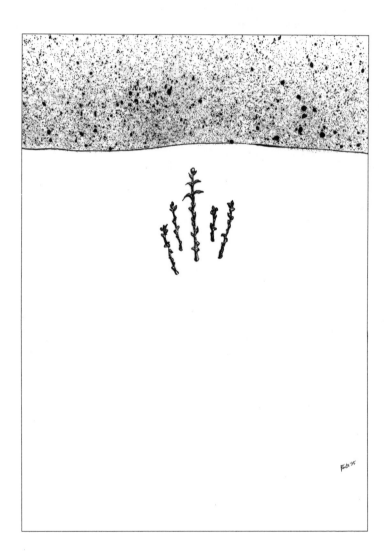

5 flora

In the gorge of Kipton Creek, the stream winds between low floodplains and steep shale cliffs. The floodplains are covered with sycamores (plane trees) and cottonwoods, ostrich ferns and scouring rushes, while the cliffs are overhung with hemlock and white pine, polypody and hay-scented fern. The difference in vegetation on the two sides of the stream is painted in color in the fall, when the reds and yellows of the deciduous floodplains meet the greens of the conifers on the steep banks across the stream. The land and its vegetation are closely related, and the walker soon learns not to expect to find a sycamore on the shale cliffs or a white pine on the floodplain.

In the Great Smoky Mountains, the Appalachian Trail winds along a high ridge forming the border between North Carolina and Tennessee. One side of the ridge, facing southeast toward North Carolina, is covered with deciduous trees, while the other, sloping northwest into Tennessee, is thick with conifers; the contrast of autumn colors is like that of Kipton

Creek on a grander scale. The hiker on the trail is met with opposing sensations when emerging from the dim light and chill of the north-facing slope into the bright warmth that faces toward the south.

At high elevations in the Smokies, one expects to find spruce and fir (where it hasn't yet been destroyed by the woolly adelgid insect) on the northern slope, and beech and red oak on the southern. In climbing from 2,000 to 6,500 feet, the hiker sees the flowers and trees of highland North Carolina (or lowland Ohio) turn into those of Maine or Canada. The plants tend to follow the same patterns up and down a mountain that they display north and south on a continent, with each 300 feet in elevation corresponding to about 1 degree of latitude. Elevation in the mountains can be estimated by noting the herbs, shrubs, and trees. When I begin to see an abundance of hobblebush in the Smokies, I reckon that I have made it to about 5,500 feet.

In the Colorado Rockies, too, different plants are at different elevations: sage and prickly pear cactus in the dry plains at 5,000 feet; ponderosa pine on the lower slopes at 7,000 feet; lodgepole pine and Englemann spruce in the greater moisture at 9,000 feet; stunted spruce and fir in the high winds and cold above that; and grasses, sedges, and alpine herbs beyond the treeline at the mountaintop. The tops of the Rockies reach, in plant life, into the Canadian Arctic. Even the lichens, which give the rocks of Colorado so much of their color, seem to know where they are growing. In ascending the mountains above the Poudre River in northern Colorado, I have found that the lichen *Rhizocarpon* gradually increases in prevalence; when 90 percent of the lichenized rocks have *Rhizocarpon* on them, I can be pretty sure that I have reached 9,000 feet. So in the eastern or the western mountains, as along the streams of northern Ohio, a close relationship exists between the nature of the place and the kinds of plants that it contains.

As each place on earth has a unique latitude and longitude, so it is also unique in the way that physical and biological coordinates come together. Across eastern North America, lines denoting average temperature and precipitation run, like latitude and longitude, from warmer south to cooler north, and from wetter east to dryer west. While elevation and proximity to large bodies of water and differences in air movements create a much more complex pattern than that of a simple grid, in a general way a given region represents a unique intersection of moisture and temperature gradients.

So too is every place a unique intersection of floral species. The yellow birch is a graceful tree whose strong wood makes a fine walking stick for a hike in the Appalachians, but in northern Ohio it seems to occur only east of the Black River, and not to grow farther west. The bur oak occurs in our area, but it is more a western than an eastern tree, being especially prevalent where the eastern woodlands give way to the western grasslands. We have lots of tulip trees, but we are near the northern limit of their distribution, as they give out just across the lake in southern Ontario. The white pine is near its southern lowland limit in northern Ohio, and we should go north or to the Appalachians for greater numbers of this tree, whose soft, strong wood the New Englanders missed when they came to this country. Only in Lorain County would we be likely to find the particular blend of yellow birch, bur oak, tulip tree, and white pine that we have here—along with the beech, sugar maple, red maple, and white ash that are among our most numerous species. There is a close relationship between the geographic location of a land and the flora it can be expected to have.

It is easy to begin to feel that in nature, undisturbed by human influences, living things are where they are because that is where they are supposed to be—in the sense that their presence represents a unique harmony between the organism and its environment. In eighteenth-century England, this harmony was viewed as the direct work of God, supporting the idea of natural theology—that God could be known through his natural works as well as through revealed scriptures. Darwin, the theology graduate, was permitted by his family to embark on the *Beagle* partly because it was believed that the study of nature would be a good thing for a future minister. Many thought that plants and animals are found exactly where God had specially created them to be. In traveling around the world on the *Beagle*, Darwin began to think that nature is a historical process, in which the present is to be understood by looking to the past.

In northern Maine and Wisconsin, I have seen the land covered, mile after mile, with nearly pure stands of white birch or aspen, while areas not far away were rich in their mixture of trees. In such places, the hand of history lies on the land, history perhaps of human making, but often of nature's own. Large stands of pure birch or aspen indicate that a fire swept through the region not too many years before. The roots of aspens surviving a fire grow outward rapidly, sprouting new stems that thrive in the sunlight of an open area. Birch has fine seeds that the wind distributes widely and quickly from surviving trees. The land is what it is not just because of present environmental conditions, but because certain things have happened in the past. The glaciation of North America and of Europe and Asia had an enormous effect on the distribution of plants and animals on those continents. Every time a tree falls in a forest, the sunlight streaming through the open canopy changes the conditions and the life of a previously shaded spot.

Darwin saw much evidence, in his five years of traveling on the *Beagle,* that history had influenced the distribution of plants and animals in the world. The ability of an organism to survive in a particular place is a necessary prerequisite to finding it there, but the accessibility of the place to the organism is also a prerequisite. To be there, a plant or an animal must be able to get there—if not at present, then sometime in the past. Instead of a single act of creation, Darwin saw in nature a long historical venture, with species evolving in time and spreading in space, but faced with barriers such as oceans and mountains and deserts, which prevented their reaching all suitable habitats in the world. Darwin visited the Cape Verde Islands in the Atlantic off Africa, and the Galápagos Islands in the Pacific off South America. The islands are similar in two important ways: both are volcanic in origin, and both lie near the equator. Yet the plants and animals of the Galápagos differ greatly from those of the Cape Verde Islands. The life of each island group resembles that of the continent from which, through history, it might most easily have come.

Throughout much of the evolution of the flowering plants in the Cretaceous and Tertiary periods, North America was joined on its east with Greenland and Europe, and on its west with Asia. Judging from the fossils from those times, this entire northern supercontinent enjoyed a warm, temperate climate even in places (such as Greenland and northern Alaska) that are now very cold. The bald cypress genus, *Taxodium*, which grows today at 41 degrees north latitude in Ohio's continental climate, or at 51 degrees in the English oceanic climate, was growing then in a region that is now (as a result of drifting continents) at 81 degrees north latitude. A common flora tended to spread across the whole northern world. During the early part of the Tertiary period, an inland sea east of the Urals—the Straight of Turgai—created a barrier to the migration of plants and animals, so that at that time North America was central to Europe and Asia, which were otherwise isolated from each other. Later, the rising of the Rocky Mountains created a barrier in North America and a heterogeneity in the pattern of rainfall across the continent; the dry region east of the mountains disrupted the continuity of the northern flora.

The flora of eastern North America remained very similar to that of Eurasia for a long time, and because of the stability of many plant groups, is still so today. Many species are identical, or nearly so, and where species vary, the genus to which they belong may be shared. Our American beech (*Fagus grandifolia*), is almost identical to the English (or European) beech (*Fagus sylvatica*). Walks in the Ohio beech–maple woods have made me feel very much at home in the beech woods of southern England.

In the countryside of Kent, near Down House where Darwin lived, are many public footpaths through the fields and woods. One beautiful June day, I set out from a little railway station in Kent to find my way on these footpaths to Darwin's house—to see the field maples and English oaks, the hawthorne and the elders as Darwin might have seen them. On a little hillside, I found myself walking past clumps of a robust fern that looked much like our North American bracken, so I inquired of an Englishman coming the other way, "What do you call this fern that looks so much like bracken?" He answered, with reserved amusement, "We call it bracken." Botanists on both sides of the Atlantic call it *Pteridium aquilinum*. In the plant world, so many of the friends we make at

116

home will be there to greet us in some other part of the Northern Hemisphere.

When Darwin was walking the public footpaths around his home, great minds around the world were struggling to understand the role of history and environment in the geography of plants and animals. From Borneo in 1855, Alfred Russel Wallace wrote an essay entitled "On the Law which has regulated the Introduction of New Species," in which biogeography and history were brought together in a global theory for the first time. The law, in Wallace's words, is that "every species has come into existence coincident both in time and space with a pre-existing closely allied species." For three years in the East Indies, Wallace pondered how species could gradually change from one form to another, while Darwin struggled with the same problem as he walked his "thinking path" in the little woods behind his home. In 1858, they both came to the same conclusion: that prolific reproduction combined with a natural selection of variable offspring could give rise to gradual change, and the origin of new species.

In 1858, Louis Agassiz was teaching at Harvard, and so was American botanist Asa Gray. Gray had been comparing botanical specimens recently received from Japan and China with plants collected in western Europe and eastern North America. It must have been shocking for Bostonians to learn from Gray that the exotic Far East, so distant and so recently opened to Western eyes, had plants very similar to those of New England, more similar even than those of Europe. The tulip tree, sassafras, and hickories, for example, which were absent from Europe, were present in Asia.

Gray thought that the floral similarity of eastern Asia and eastern North America must have a historical explanation, and he guessed that Agassiz's glaciers were an important part of it.

But Agassiz, while he delighted in interpreting the physical landscape in historical terms, with emphasis on the role of glaciation, was unwilling to apply historical contingencies to the plants and animals of the land. He believed that living things were created by God to occupy their present places. The man who brought such a revolution to geology, and created the ideas for explaining so much of the present geography of life, refused to interpret biogeography in historical terms or to accept the evolutionary view of Darwin and Wallace. Gray, however, embraced evolution and remained convinced that global but natural processes must be behind the present distribution of plants and animals on the earth.

There was another Bostonian in 1858 who saw history at work throughout the natural world and who could not accept Agassiz's idea that the distribution of plants and animals was to be explained by special acts of God. Henry David Thoreau did not have to travel to the far corners of the earth to find global principles in nature. In the woods and fields near his home in Concord, he could see how the squirrels carried pine cones around, and how the wind blew vast numbers of birch seed across the winter snow. He could trace the presence of young pines in a meadow to the parent trees from which they had come, and note how the growth of the new pines began to change the grasslands into woods. In the dispersion of seeds, and in their germination and growth, Thoreau saw a dynamic, creative nature, ever turning the landscape into something new. He concluded that creation was not a single act in time, but that all time is involved in creation.

A great many native Ohio plants have close Asian cousins. As summer comes to my garden, phlox appears in great numbers. Our garden phlox is derived from indigenous American and Asian species, and all the phlox grown in Europe has come from them too. *Phlox* is one of many genera of plants that are shared by eastern North America and eastern Asia, but that are not native to Europe. Other such genera include, among the woody plants, *Magnolia*, *Liriodendron* (tulip tree), *Liquidambar* (sweet gum), *Nyssa* (sour gum), *Catalpa*, *Sassafras*, *Hamamelis* (witch hazel), *Tsuga* (hemlock), *Carya* (hickory), and *Celastrus* (climbing bittersweet). Among wildflowers that are shared are *Trillium*, *Podophyllum* (May apple), *Symplocarpus* (skunk cabbage), and *Jeffersonia* (twinleaf). Likewise, the sensitive fern (*Onoclea sensibilis*) and cinnamon fern (*Osmunda cinnamomea*) are found in eastern Asia and America, but not in Europe.

Many of the trees and wildflowers that are common to North America and Asia, but absent from recent Europe, have been discovered in Tertiary fossils in Europe. Among those found in European preglacial deposits are the hemlock, hickory, sour gum, and tulip tree.

Magnificent tulip trees, 6 or 7 feet in diameter, can be found in the lower elevations of the Great Smoky Mountains. The tulip is the tallest, straightest tree in the Ohio woods. In spite of its grandeur, it is easy for an Ohioan to begin to take it for granted. But I was strolling through the back garden of Gilbert White's idyllic Selborne home one day, delighting in the affectionate ways that wild nature and humanity had been brought together there, when a British tourist pointed excitedly to the tree overhead, the principal object of his interest: a large American tulip tree. I was pleased that this common Ohio tree was given such meaning in such a special place. It seems fitting that a tree once native to Britain, but driven from the island by nature's own history, should be restored to its ancient home by human care.

A back garden can preserve plants that once were very rare.

The dawn redwood tree, thought by Western botanists to have gone extinct until it was found in the twentieth century in the mountains of China, now grows in gardens around the world. Gardens also create hybrids not previously known. The plane tree ("sycamore" in America) was lost to Britain in the Ice Ages. In the seventeenth century, the American and Asiatic plane trees were brought to the Oxford Botanic Garden. The trees hybridized, creating a new European tree called the London plane tree. The parks of London are full of them, as are the parks of many American towns.

The distribution of native trees and wildflowers in North America, Europe, and Asia, together with the fossil record from the Tertiary period, suggest that a common flora once spread across the entire temperate region of the Northern Hemisphere, but that many of the plant species were driven out of Europe during the glaciations, so that the flora of Europe became somewhat impoverished compared with that of North America and Asia (and European botanists and gardeners of the sixteenth to nineteenth centuries could be excited by the new plants discovered in those far corners of the world). In 1915, Eleanor and Clement Reid suggested that the glaciations were harder on the flora of Europe than on that of North America and Asia because the mountains of Europe (the Alps and Pyrenees) lie east and west, while those of North America and northeastern Asia trend more north and south. As glaciers advanced from the north, temperate plant species could continue to reproduce toward the south in America and Asia, but not in Europe, where mountain glaciers blocked a southerly retreat. Many species found no refuge in Europe from the cold climate and were extinguished. In Asia and North America, plants could reach refugia in the south, from which they advanced northward again when the ice melted.

During the glaciation of northern Ohio, all plant life was driven out, but found refuge farther south. Some of our present species may have survived in southern Ohio, others in the southern Appalachians, and still others along the coast of the Gulf of Mexico. The climate of southern Ohio may not have been very severe, even when an ice mass several thousand feet thick covered northern Ohio. Ice is pushed south by accumulations much farther north, and its southern edge occurs where melting is fast enough to equal the rate of delivery. As the botanist E. Lucy Braun, put it: "The position of the ice margin is determined by the warmth of the climate, not by its coldness." On the coast of Greenland, at latitudes of 60 to 80 degrees, 400 plant species representing 50 families are able to grow just 40 miles away from ice thousands of feet thick. Nevertheless, in Europe many species could not survive in the region between the Scandinavian and Alpine glaciers, and it is probable that most of the present northern Ohio species took refuge far south of the Ohio River.

The postglacial history of plants in North America and Eurasia has been studied by analysis of pollen and other plant fragments that have accumulated in bogs and lake-bottom sediments. Bogs, with their high levels of acidity and low levels of oxygen, have served as natural museums, preserving everything from dug-out canoes to human bodies. Pollen preserved at different levels in accumulating peat can indicate the flora and climate of a region at different postglacial times. The glacier, in leaving "kettle-holes" for ponds and bogs and in destroying previous drainage patterns so that numerous lakes were formed, provided many collection points for pollen blowing in from grasses, herbs, and trees.

Anyone who has parked a car under a pine tree in early June, or has suffered with hay fever, will know that there can be considerable pollen in the air. Pollen can be identified, even after thousands of years, to the genus and sometimes to the species of plant, and the relative frequency of a given pollen gives some idea of the abundance of that species at a particular time (determined by radiocarbon analyses) for the region from which the sample was taken.

Twenty-two bogs in northern Ohio were sampled and their tree pollen analyzed by Paul Sears and Loren Potter in the 1940s. The oldest (deepest) peat layers had pollen from northern, cold-adapted evergreen trees, while the higher layers had pollen from more southern, deciduous trees. The order of average first appearance in the layers was spruce, fir, pine, birch, oak, hemlock, hickory, and beech. Potter pointed out that temperature alone did not determine the order of appearance, for hemlock is a slow migrator, and beech requires a good soil.

More recently, Margaret Davis has used radiocarbon-dating techniques as well as pollen analyses in studying postglacial sediments from many different sites in North America, and has been able thereby to map the migration of plants northward following the retreat of the glacier. Because the southern Appalachians have such a richly diverse flora, exhibiting at different elevations so many of the species of the eastern United States and southern Canada, it had been thought that they were a refuge for all the species they now contain, and served as a center from which all those species could begin a migration northward. It is now thought that even the beautiful Smokies suffered some loss during the glaciation, and that maples and chestnuts had to migrate back there from a refuge along the Gulf coast.

Hemlock and white pine probably returned to northern Ohio from the Appalachians, but maples and chestnut from the Gulf. Hemlock seems to have lagged about 1,000 years behind the white pine in its migration, perhaps because it requires a richer soil than the pine. Today along the steep sides of the river gorges, hemlock and white pine stand together, but the pine may have had to wait 1,000 years before it was joined by the other tall conifer. Davis has estimated that the white pine and white spruce migrated northward at about 300 meters a year, the hemlock and maple at about 200 meters a year, and the chestnut (whose seeds require squirrels or other animals, rather than the wind, to distribute them) at only 100 meters a year.

As it recovered from the glaciation, northern Ohio saw flora from all the elevations of the present Appalachian Mountains pass through its land. At dawn on an Easter morning, I put on my frozen boots and left the icy shelter atop Mount LeConte, to descend through early morning sunshine and the awakening of an Appalachian spring. I felt that I was witnessing millennia of postglacial history as I descended from the spruce and fir in the melting snows at the top, past pines and galax and trailing arbutus on rocky outcrops 1,000 feet below, and down through the darkness of great hemlocks to the hepaticas, trout lilies, and trilliums unfolding in the sunshine of the hardwoods at Cherokee Orchard. In one morning, I seemed to have walked from northern Maine to upland North Carolina, and from ancient to present-day Ohio.

In northern Maine, the glacier left behind a frozen ground, which became tundra for some time before the soil thawed and the conifer trees were able to move in. In the milder Great Lakes area, spruce forests were able to establish themselves quickly, and there was no extended tundra period after the ice left. Grasses grew as well (producing enough pollen to be found 10,000 years later), and so the forest was probably an open one. Mastodons browsing on the spruce and tamarack could move freely through the forest, as could the newly arrived Paleo-American hunter.

About 8000 B.C., the entire Great Lakes area saw a sudden decline in spruce; pine, birch, and oaks replaced it, with hemlock waiting for its chance on better soil. The woods became more closed, and the mastodon disappeared into more open areas and then into extinction. The trees of the Far North gave way to those adapted to a somewhat warmer climate. Some trees (for example, the spruce, fir, tamarack, birch, and aspen) can withstand the earth's coldest winter air. In winter, they pull the water from their living cells into extracellular spaces where it can freeze without injuring the cells. Other trees (oak, ash, and elm) cannot perform this osmotic trick, and survive only to about -40 degrees (Fahrenheit or Centigrade—they are the same at that point) by supercooling the water inside their living cells. If the winters are harsh enough, only the first group of trees can survive. But in somewhat milder climates, the second group can also withstand the winters, and may be able to make quicker use of summer's warmth and water.

By 5000 B.C., the climate of northern Ohio had warmed so much that it was actually warmer than it is today. It may also have been drier. Asa Gray, in 1878, was the first American botanist to point out that the grasslands of the West are drier than the forests of the East. If the amount of rain and snow has changed over the centuries, then the distribution of forests and grasslands may have changed too. In 1902, Charles C. Adams proposed that the grasslands of the West had once, in postglacial times, extended much farther east than they do today. He called this eastern extension of the western grasslands the "prairie peninsula." The idea that prairies extended into Kentucky and Ohio, and perhaps beyond, during a postglacial xerothermic (dry and warm) time, was subsequently championed by Henry A. Gleason of the New York Botanical Garden, Edgar N. Transeau of Ohio State University, and E. Lucy Braun of the University of Cincinnati. Braun thought that there had been prairie extensions at two different times: before the Illinoisan Glaciation as well as after the Wisconsin Glaciation. A number of isolated prairie patches in Ohio survived until the nineteenth century, as did extensive grasslands in western Kentucky. A northern Ohio high-grass prairie has been restored at Killdeer Plains in Wyandot County.

Whether or not extensive grasslands spread across northern Ohio during the warm climatic optimum, the woodlands probably became more open, especially on the sandy beach ridges. Prickly pear cactus probably grew in the dry sandy soils of the beach ridges; it survives today at Point Pelee on the Canadian side of Lake Erie.

By the time of the climatic optimum in the southern Great Lakes, enough glacial meltwater had returned to the world's oceans that their coastlines approached those of today. The Bering region uniting Asia and North America had become flooded 8,000 years earlier. But now the North Sea and Baltic Sea separated Scandinavia from Britain, France, and Germany; the Thames became a river solely of Britain and not a tributary of the Rhine; and the British Isles became divided by the Irish Sea and separated from the rest of Europe by the flooding of the English Channel.

As the ocean waters rose, Scotland provided the last bridge (a northern one) between England and Ireland. Some plants and animals that arrived back in England shortly before the English Channel formed were unable to reach Ireland through the chilly northern bridge. (Britain did not see its own climatic optimum until about 2,000 years after ours, in 3000 B.C.) The linden (or lime, or *Tilia*, or—as we call it here—the basswood)

and hornbeam were trees that reached England after the glaciation, but were unable to get to Ireland—until people took
them there. It was the glacier rather than St. Patrick that drove
the snakes from Ireland, for like the linden and hornbeam
trees, snakes could not find a warm enough route back to
Ireland. Britain is full of trees today that did not return before
the Channel formed, but were reintroduced by people (for
example, chestnut, sycamore maple, Norway spruce, European
larch, and fir). The only gymnosperms (trees with naked seeds)
that made the postglacial trip back to Britain without human
help were the Scot's pine, juniper, and yew. Many of the beautiful stands of evergreens ringing the mountain tarns in
Wordsworth's Lake District are made up entirely of trees not
native to postglacial Britain.

While no physical barrier prevented the migration of species
back to northern Ohio following the glaciation, sheer distance
and lack of time may have done so. There are several trees from
farther south that people have brought to northern Ohio and
that seem to flourish in this climate. Perhaps those trees would
have arrived here by their own seed-dispersal devices if they
had had more time—or perhaps they would have found a biological barrier (such as the trees that were already here) that
prevented their reproducing themselves in our woods. Two of
these southern trees were introduced to Ohio farms as a means
for enclosing fields.

Behind my house is a grove of tall catalpa trees spaced at
regular intervals, a reminder that my home was once a farm on
the outskirts of the village. The trees were planted about eighty
years ago to grow a crop of fence posts, but the crop was never
harvested. Catalpa, like many other trees, will sprout from its
stump when cut, quickly producing a new crop of poles or
posts. Its wood is soft, so wire can easily be stapled to it, yet it

126

is highly resistant to rot and will last for years even on wet glacial clay. In England, periodic cutting (or coppicing) is a common method of managing woodlands, and coppices of hazel and other trees are a charming part of the English landscape. The growth of wood is rapid in a coppice, because the large root systems are already established to draw water and minerals for the new, young shoots. The catalpa of my old farm is native to the Mississippi and lower Ohio River valleys. It reproduces itself in my yard, but I have seen no evidence that it has taken to our native woods; in that sense, it may not yet be a truly naturalized citizen of this land.

Along some of the roads of our countryside are lines of osage orange trees with their shiny leaves, large green fruits, and thorns. They were planted close together as living hedges, their thorny branches acting as a natural barbed fence to keep cattle in or out of a pasture or field. In England, the hawthorne has been used for thousands of years for the same purpose. The osage orange is a native only of a small area of Texas, Oklahoma, and Arkansas, yet it thrives in northern Ohio. Its heavy fruits seem ill-adapted for wide dispersal by our present fauna, so without any other discernible barrier it has remained in a restricted range except as people have brought it to their farms and homes.

The osage orange hedge was the Ohio equivalent of the stone wall of rural New Hampshire and Vermont. Just as stone walls can be found in New England woods where old fields have been abandoned and allowed to grow up to woods, so a line of osage orange trees or their stumps can sometimes be found in the woods of northern Ohio. The hard wood of osage orange contrasts with the soft catalpa, but like catalpa, it is resistant to decay, and stumps of osage orange can mark an old field boundary long after the field itself has vanished.

The sweetgum and the bald cypress are two more trees whose seeds might have been dispersed naturally to northern Ohio if they had been given more time. They are trees of the towns rather than the farms, having been planted more for aesthetic than practical purposes. The star-shaped leaves of the sweetgum turn deep red in the fall, and the graceful, pendulous fruits look like Christmas ornaments. The sweetgum's native range had reached as far as southern Ohio before people brought it north across the state. The bald cypress is a native of the Mississippi and Ohio valleys as far as southern Illinois and Indiana, and of the Atlantic coast to Maryland, but it is such a majestic tree that it has traveled to gardens and parks around the world. There are magnificent specimens, for example, in the University Parks in Oxford and on Tappan Square in Oberlin.

A few plants returning to our area found a unique locale at Camden Bog. This is a peculiar topographic spot, a pond partly filling a depression in the land, with little inflow and no outflow. How can such a thing be formed by normal erosion? It is as though somebody had dropped something very big in the soft glacial mud, and then spirited it away, to leave only a puddle of water. I would like to know what legends surrounded it in the stories of Native Americans. Many such ponds across the continent exist where the ice once melted, and the explanation of geologists is that they are the result of huge, terrestrial icebergs that once broke off the melting ice sheet. A block of ice was buried at Camden by silt and stones washing out of the melting glacier. The ice preserved a depression in the glacial outwash, and as it slowly melted its cold waters formed an isolated "kettle lake." Around this lake a number of northern plants were able to survive as the climate warmed.

One of these northern plants was sphagnum moss. Sphagnum produces acidic compounds that leach into the water, creating an acid environment in which the roots of most plants find it difficult to absorb water. The growth of vascular plants is therefore slow and limited to species that can survive such acidic conditions. Decay of vegetation is also slow, as microbes are inhibited by the acidity and the lack of oxygen in the stagnant waters. The sphagnum thrives, creating a thick blanket that retains the coolness and humidity of the lake, but in which few plants can take root.

Swamp loosestrife is one plant that is able to root in sphagnum. It can also invade the edges of the water by landing on it rather like an amphibious airplane: When its drooping branches touch the water, they form spongy, floating tissue, which buoys further growth. From the water's edge, the swamp loosestrife and sphagnum can form a raft of plant and animal life floating on the water. Reinforced by leatherleaf, sedges, cranberries, and the insect-eating pitcher plants and sundews, the mat can achieve sufficient strength and buoyancy to support a somewhat tottering person. Such a floating mat is often called a "quaking bog."

Camden Bog once had such a floating mat. It was destroyed in the 1950s when the city of Oberlin several times pumped the lake nearly dry for its water. The mat was pinned down by falling trees, and at least thirty-eight species of plants were lost. But in recent years, six of those species have returned, probably from surviving seeds. Sphagnum moss is still present at the edge of the pond, and swamp loosestrife is growing with abandon, so it is possible that a new floating mat will form. Buttonbush, winterberry, and marsh fern give testimony to the bog's unique place as a relic of our glacial past.

Ecology intersects with history in determining the flora of

any region. Since life is made of water more than all other things combined, the greatest environmental factor for life is water. If organisms do not actually live in water, they must have a source for its uptake and transport to all their living cells; a few desert animals find all their water in their food, but their source of food had a source of water. There is no life without it. Much of the effect of temperature on life is due to its influence on the availability of or requirement for water. Water is not usually available if it is frozen (though skunk cabbage in spring produces enough heat to melt it), so a continental ice sheet is like a desert. In tundra, the growth of trees is prevented either by ground that is permanently frozen beneath the surface, so that large root systems cannot grow, or by summer temperatures that are too cool to permit adequate growth of aboveground woody tissue. Life must stay close to the summer water under such conditions, and the tundra has only grasses, sedges, lichens, herbs with their perennial parts at ground level, and a few annual herbs that winter over only as seeds.

Trees such as spruce and fir can withstand the coldest of winter temperatures, but seem to require at least eight weeks of summer temperatures averaging above 50 degrees Fahrenheit. Where July temperatures average 50 to 68 degrees, a boreal forest dominated by spruce and fir is found; with warmer July temperatures, the deciduous woods takes over. Oberlin's average July temperature is about 72 degrees Fahrenheit, so it is well within the deciduous forest, and neither spruce nor fir is native to northern Ohio. White pine, hemlock, red cedar, common juniper, and Canadian yew are the only gymnosperms native to Lorain County, and they are found in small numbers. Tamarack is native to Cuyahoga County to the east, and northern white cedar to Erie County to the west.

The deciduous forest extends from the Great Lakes region south to the subtropical evergreen forests of Florida, where the winter temperatures are warm enough to sustain water transport through the trees year round, and it is no longer adaptive to lose the leaves (all at once) and become dormant. In the West, the deciduous woods are limited by decreasing rainfall, and grasses take over where the trees cannot grow. The trees can survive low precipitation better in the cooler North (where evaporative water losses are less) than in the warmer South, and the boundary between the eastern woodlands and the western prairies therefore runs northwestward.

Because many of the individual species of trees have limited ranges in the deciduous woodlands, and all have varying frequencies of occurrence, the deciduous forest changes character from one region to another. Lucy Braun was the first to try to describe these changes; she traveled over 60,000 miles through the eastern woodlands mapping the whole into different regions, based on patterns of association of trees. Where regions had been largely cut over (as in northern Ohio), she consulted records of the surveyors who saw the land before it was deforested.

No one species takes over the forest canopy in any region, but mixtures of two or more associated species might be dominant, and therefore useful for characterizing an area. Braun characterized the woods of our land as part of a beech–maple forest covering the glaciated region of Ohio and Indiana, and southern Michigan. The most numerous large trees in this forest are the American beech and the sugar maple. To the northwest of our area, in parts of Wisconsin and Minnesota, the beech–maple association becomes a basswood–maple forest, where the basswood takes over for the beech, which is absent altogether in the western part of the deciduous woods.

The beech and sugar maple are found on moist but moderately well-drained soils. One can say that northern Ohio is (or was once) a beech–maple forest, but actually it is a patchwork of different tree associations. Soils vary from clay over much of the region, to sand on the beach ridges and on parts of the floodplains of the streams. Even on a relatively flat landscape, the surface drainage varies a great deal, and near our streams the slopes become pronounced, affecting not only drainage, but also exposure to the sun and winds. One place becomes many places. Oak and hickory dominate the drier parts of the Ohio beech–maple woods—as on well-drained slopes close to the high edges of the river gorges. Ash and red maple dominate many of the wet places where, until the advent of the Dutch elm disease in the 1960s, they were associated with the American elm.

Red maple and ash, with their winged seeds cast in great numbers, their broad tolerances for soil, and their fast growth in bright sunlight, are also good at taking over abandoned fields, unweeded flower beds, or uncut fencerows.

The white pine requires substantial sunlight for its growth. Its graceful, open appearance results from the fact that its branches are careful not to shade one another. A mature tree can keep reaching for the sky, dropping its lower branches as it goes, but the young trees need fairly open sites. This they find mostly along the steep slopes of our streams, where there is a ribbon of open sky uncrowded by deciduous trees. The hemlock is there too, but in the shadier, northern exposures; its small seeds may have difficulty rooting in the leaf litter of the deciduous woods.

The plane tree (sycamore) grows mostly on the floodplains of the streams, perhaps because its large fruits are easily carried by water. Cottonwood, red maple, and butternut also favor the wet conditions of the floodplains. The sandy beach ridges had many American chestnuts until the chestnut blight removed them.

If different trees prosper in different circumstances of soil, moisture, and sunlight, the trees themselves affect the ecology of the woods in different ways. Sometimes these effects are very specific: It is only under beech trees that beechdrops grow, for this little flowering plant has scale-like leaves devoid of chlorophyll and must tap into the roots of beech trees for its nutrition. Often the effects are more general: Trees reduce (to different degrees, depending on the species) the available sunlight and thus affect a broad range of plants and animals. There is little grass under the beech tree in my yard. The beech and sugar maple cast such heavy shade that many plants cannot grow while the trees are fully in leaf. The season of greatest change in a beech–maple forest is the early spring, when hepaticas, spring beauties, trilliums, and scores of other wildflowers enjoy the bright sunlight that soon will be shut out by the growing foliage of the trees.

The Ohio beech–maple forest is called a "climax forest" to indicate that it is the result of a succession of ecological stages leading to it and that it is capable of sustaining itself indefinitely, barring major disasters (such as large fires) or changes in climate. A climax forest can sustain itself only if the dominant trees are capable of reproducing themselves. Because seedlings of the beech and maple are shade-tolerant (unlike seedlings of white pine), they can grow up in the shade of their parents and maintain the same forest canopy for many years. Likewise, all components of a climax forest must be able to reproduce themselves under the conditions that they have collectively created. The detailed pattern within any forest does, of course, change, as no part is quite the same from year to year or even from day to day, but the overall composition may remain fairly stable.

Nature or humanity can have, nevertheless, dramatic effects on a forest. Climate has been changing gradually throughout the postglacial period, and species are always adapting to these changes. Two opposing natural trends probably affected the woods of northern Ohio during the past 7,000 years, following the climatic optimum. While many species were still migrating north, reclaiming land they had lost during the glaciation, other species experienced a southward retrenchment as the climate gradually cooled.

Natural fire has always affected forests from time to time, especially in areas such as northern Ohio where lightning storms are frequent. Natural population explosions of insects or other plant pests can devastate a forest. About 2800 B.C., there was a rapid decline in the number of hemlock trees throughout eastern North America, probably as the result of a destructive insect or a hemlock disease. England saw a dramatic decline in its elm trees about 3,000 B.C. Since this was coincident with the introduction of agriculture and domestic cattle into England, and since elm leaves are browsed by cattle as well as by deer, archeologists suggested that the elm was cut excessively by early farmers for cattle feed. It now seems more likely that the trees died of an ancient epidemic of the Dutch elm disease.

When agriculture came to northern Ohio about 500 B.C., people began to have an important effect on the beech–maple forest. For thousands of years, trees had been cut to provide firewood and to make openings for villages, but now the village areas were extended to include open fields for crops. With the introduction of maize about A.D. 500, populations probably increased and fields became still more extensive. The sites of villages were no doubt moved when fields became less productive or firewood gave out. We may suppose that in 1600 land use in Ohio was similar to that in upstate New York and southern New England. The Iroquois capital of Onondaga, near Syracuse, was situated in nine different places between 1610 and 1780—hence the site of the town changed about every twenty years. Likewise, the Ohio land probably exhibited a pattern of ecological changes, with new cutting and burning in some parts and reversion of old fields to woods in other places. Such a variegated land of fields and woods would have helped to maintain the populations of deer and bear on which the people depended for skins and for much of their winter food.

The many fields and villages of the Native Americans, and the open woods that they created by burning, were seen by the French in the St. Lawrence River valley and upstate New York, by the English in Massachusetts and Virginia, and by the Spanish in the southern states. Ohio must have been much the same, but no pen recorded its condition. By the time Daniel Boone and other English-Americans penetrated west of the Appalachians in the mid-eighteenth century, the country had been depopulated by disease and warfare, and much of the open country had grown back to thick forest. Nevertheless, there were many open grasslands west of the mountains. In West Virginia, George Washington owned land on the Kanawha River containing meadows abandoned by the Native people. Western Kentucky, Indiana, and Ohio had extensive grasslands deriving ultimately perhaps from the dry climatic optimum, but kept open in large measure by the Native Americans. The native Kentucky grasslands were taken over by the

European grass *Poa pratensis*, to become the bluegrass country. By the end of the eighteenth century, most of northern Ohio's surviving Native people had retreated to the Maumee River valley, where Anthony Wayne's army found open land for 100 miles along the river from Fort Wayne to Lake Erie. The rest of our land had nearly 200 years to surrender its open fields and woods to a thick beech–maple forest, into which the New England surveyors and settlers came.

It took less than a century for the 7,000-year-old forest of northern Ohio to be converted to open land by the New England settlers. In 1853, more than 40 percent of the forest was gone, and by 1940 about 93 percent. A survey in 1940 recorded 170 wooded areas in Lorain County comprising 24,000 acres; 47 percent of the forest remnants were classified as beech–maple woods, 39 percent as elm–ash–red maple, 10 percent as oak–hickory, and 4 percent as floodplain (sycamore–cottonwood–red maple) woods.

The woods were replaced not only by agricultural crops, but by a multitude of plants adapted to rapid growth in bright sunlight and in disturbed soil. Many of the most successful of these were not indigenous American plants, but Eurasian weeds that had come with the English settlers. As early as 1638, a visitor from England noted twenty-two familiar European plants growing wild in Massachusetts, including dandelions, plantains, nettles, and dock. The Native Americans in Massachusetts and Virginia called the broad-leaved plantain the "Englishman's foot," recognizing that it seemed to go wherever he went. Its seeds may well have been in the mud of his boots, as well as mixed with the seeds of his agricultural crops.

In country and town, wherever the sun shines brightly, the ground is covered today with Eurasian plants. Pastures are filled with European grasses, invaded by and interwoven with white clover, red clover, yarrow, Queen Anne's lace, sheep sorrel and dock, St. Johnswort, buttercup, and many other herbs from the Old World. Native American plants are present too— such as fleabane, strawberry, beardtongue, goldenrod, and aster, but the impression is on the land that the king, rather than the colonists, must have won the War of Independence. America's fields fly the flags of Europe.

The arable fields of northern Ohio are filled mainly with American maize and Eurasian soybeans, but they contend with a host of Eurasian weeds, from winter cress in the early spring to many other mustards, sow thistle, shepherd's purse, velvet leaf, wild buckwheat, lambsquarters, bindweed, and jimsonweed. Even the Canada thistle is really from Europe. The native ragweed, pokeweed, ironweed, and so forth do very well themselves, but seem less successful at taking command of the fields than do their European allies.

Along our roadsides are daisies, yellow birdfoot trefoil, pink crown vetch, sweetclover, teasel, burdock, and the blue flags of chicory. Our railway beds have mullein, everlasting pea, yellow goat's beard, nightshade, rough-fruited cinquefoil, and bouncing bet—all from Eurasia.

Whether these plants new to America are weeds or wildflowers depends on one's purpose or point of view. We might say that a weed is a plant that we try to get out of our way, while a wildflower is one that we go out of our way to get. Some people try to remove every broad-leaved plant from the front lawn, while I mow around them so they will have a better chance to flower and set seed. Among the Eurasian flowers in my lawn—in addition to a surplus of dandelions, plantains, and chicory—are speedwell, chickweed, hop clover, sorrel, mayweed, self-heal, moneywort, gill-over-the-ground, dead nettle, Deptford pink, and English daisy. I am especially partial to the English daisies, and am glad to see them spreading downwind from my lawns. When a man from a chemical lawn-treatment firm called the other day to propose that he get rid of all these treasures, I told him he was speaking to the wrong person!

On the college square in Oberlin, all the native trees were cut down in 1833, except for one young American elm (the "historic elm"), which stood until the 1960s near the site of Peter Pease's cabin. It was not until 1836 that the area had been cleared of all the resulting stumps. And it was only two years after that, human nature being as it is, that people began to think that it would be a good idea to plant new trees! Beginning in 1838, elms and other native trees were transplanted to the square, and in 1856 the first exotic species, a Scot's pine, was planted. Of the fifty-four species of trees on Tappan Square in recent years, twenty-five are natives of northern Ohio, eight are from other parts of North America, twenty are Eurasian, and one (the London plane tree) is a hybrid of native and Asian ancestry.

On a fine October day in 1993, I watched as a large American elm on Tappan Square was taken down with chain saws, leaving only 1 elm still surviving on the square—out of 136 that had been there in 1945. Elm trees began to die in Europe after World War I—the victims, some thought, of poison gas used in the war. But in 1921 a Dutch scientist, Marie Beatrice Schwarz, showed that a fungus, *Ceratocystis ulmi*, was associated with the disease. Using principles defined by Robert Koch for identifying an infectious agent, Schwarz showed that the elm disease developed in those trees injected with the fungus and that the same fungus could then be isolated from the affected trees. Because the nature of the disease was discovered in Holland, the world (ungratefully) has come to call it the Dutch elm disease.

The elm fungus was inadvertently imported into the United States in 1930 with logs purchased for making veneer. A European bark beetle, which had previously entered the United States about 1900, carries the fungus and infects the elm when it feeds on its leaves and bark. To a lesser degree, an American bark beetle does the same. Bark beetles deposit eggs under the bark of dead elm trees. If the heavy bark is pulled from a dead elm, centipede-like patterns left by the bark beetle usually can be seen in the underlying wood; the "body" of the "centipede" is the remains of the beetle's egg gallery, while the "legs" are the impressions left by the beetle larvae eating their way out of the gallery.

From its origins on the east coast in the 1930s, the elm disease spread west and north, decimating the trees of northern Ohio in the 1960s. There are empty places still where the elms used to be. In Europe the elms at first seemed more resistant than in America. After an epidemic in England in the 1930s, the disease largely disappeared, but it came back again in the 1970s to destroy most of the English elms of the southern counties. The deer in the deer park of Magdalen College, Oxford, no longer stand on their hind legs to reach the tender elm leaves overhead. Recent studies suggest that this disease has hit the elms of Europe many times since the elm decline of 3000 B.C. Like smallpox and measles, however, the Dutch elm disease was unknown in America until ships brought it across the Atlantic.

Another Eurasian tree disease, caused by the fungus *Endothia parasitica*, had little effect on the Eurasian chestnut, but proved devastating to the American chestnut. The chestnut blight appeared among trees in the New York City Zoo in

1904 and spread rapidly outward, destroying one of the most prevalent trees along the east coast, in the Appalachian Mountains, and along northern Ohio's beach ridges. Chestnut wood is resistant to decay, so remains of dead trees can still be found, and young trees occasionally grow back from old stumps. A lone chestnut, big enough to produce fruits, stands above the road at Swift's Hollow near the Vermilion River.

The flora of the land is always changing, as is the light upon the landscape. What we see has never been quite the same before, nor will it ever be again. The sun moves across the sky, the seasons change, and history transmutes one pattern into another. I have stood on the high edge of Black Down in England, on a path once walked by Tennyson, and looked down upon the Weald with its patchwork of fields, lanes, and hedges, and watched little trails of smoke rising from the farmhouse chimneys toward the cumulus clouds. The wind drove the clouds, like frisky colts above the down, and the sun gave their shadows to the earth, galloping across the fields. Sunlight and rain and hail took turns in speaking to the land. Had no one walked the miles from the railway station that day, that changing light and music on the pines, the bracken, and the gorse would have gone unseen and unheard—for yesterday and tomorrow would be other days. Creation does not wait for other times or places; it is, as Thoreau said, here and now.

In my backyard in Oberlin, I used to keep a field of tall grasses and weeds, to watch them change with the seasons—to see the sheep sorrel redden in spring, and the asters and goldenrod bloom as summer ended. They brought me news each day of a happier tone than that which I read in the paper. One chilly morning in early fall, with the first frost on the grass and folded clover, I went out to visit with the weeds—when my eyes were caught by a small oak seedling. Its russet leaves were edged with frost crystals grown from the clear night air. Before long, however, I saw that the light on this etching was changing, as the first rays of morning sun glittered on its ice; a few moments later the light spread, the leaves warmed, and the crystals began to melt. So has it been with time ever flowing through the flora of the land—hundreds of millions of years, summed from moments like those, generated by the soil and the sunlight and the cool night air.

6   f a u n a

Spring begins in our woods with the skunk cabbage pushing up through frozen soil, and with hepatica, toothwort, and trillium unlocking the colors of the sun. It begins in our gardens with the aconite and crocus, the scilla and the daffodil. But it is not really here until that magic day when the chipmunk emerges from its winter burrow and sits on the steps of my back porch.

I wonder what a chipmunk thinks of the new spring world outside its burrow. It seems attuned to whatever it finds, however long its absence. Its happiness with the day, basking and scurrying, brings to mind one of the deep mysteries of life: What is the source of the knowledge and harmony that creatures bring to their world, inherited from a distant past? The chipmunk's recognition of its home in nature is part of the stream of memory passed down through every animal species. Plato wondered about it long ago and decided that knowledge is remembered from a previous life; biologists today can make it much more complicated, but Plato had the essential idea.

Life and its world have evolved together, conversing to each other through millions of journeys around the sun and through multitudes of generations.

The life cycle of any animal is too incredible for any of us to understand. How can a monarch butterfly drop an egg on a milkweed leaf and expect any good to come of it? How can a solitary monarch larva know how to create its own pupal case? How can the emergent butterfly know, when late summer comes, that it should fly off to the mountains of Mexico for the winter? A few weeks before, it was a lonely egg on a milkweed plant, with no parents or siblings to guide it. Where in the egg are the mountains of Mexico?

I think my favorite story of this kind is that of the Atlantic shearwater. The shearwater is an oceanic bird. It flies above the waves, often shearing the water with its dragging feet, as it hunts for its food in the sea. When it rests, it rests on the water and comes to land only to reproduce. At nesting time, it flies thousands of miles to the South Atlantic, to islands off Antarctica where no predators will disturb its young. The parents dig out a burrow on the cliffs of the island, and the mother lays a single egg. When the fledgling hatches, the parents feed it with regurgitated food they have captured from the sea. The young one grows so well on this diet that it becomes fatter than its parents and may be unable to leave the burrow. Whether in disgust or frustration, or more likely in quiet wisdom, the parents have the good sense to leave—to abandon the young one to its own devices while they fly back to the North Atlantic. The young shearwater starves—so what have the parents accomplished from all their trouble? In starving, the young bird becomes slimmer, and its flight feathers mature, so it can now exit from the burrow and exercise its wings. It has no brothers or sisters, no parents to tell it where to go, but one day flaps its wings and flies from its island birthplace—to the North Atlantic, to find the same foods its parents hunted and live the same life they knew. It is a little too simple to say that it is all in its DNA. But then, where is it?

The chipmunks in my yard have their burrows under the granite boulders (probably glacial erratics) that were used to face the sandstone foundation of my home. They do not seem to mind that the rocks have been moved from the native woodland where their ancestors lived long ago. The generations of chipmunks have seen in the granite rocks one of the essential elements of their lives and have followed them from the woods to the edges of my house. As the world has changed for succeeding generations of chipmunks, each waking to a different scene, they have managed to adapt the memories of the species to the new life before them. Many animals, of course, have been unable to adapt to the changes that nature or humanity have wrought, while others have found, like the chipmunk on my back porch, new opportunities open to them.

The chipmunk is one of about forty-two mammals living wild in the Oberlin region today (I have not counted people!—or their domestic mammals, or feral dogs or cats). Of that number, forty are natives that have been here far longer than people have, while two (the house mouse and Norwegian rat) came with the European settlement. Our mammalian species are only a tiny sample (about 1 percent) of the world's total, but they represent seven of the world's nineteen mammalian orders; until a few thousand years ago, northern Ohio could claim two further orders (wild horses and elephants). Many animal enthusiasts do not like to admit that nearly two-thirds of the world's furry species are either bats or rodents; that proportion holds for Ohio too, but those of us who like chipmunks can hardly object.

145

About a dozen of the natives I have seen in my yard, and half a dozen have made it into the walls or rooms of my house: the white-footed mouse, shorttail shrew, little brown bat, and red, gray, and flying squirrels. A young flying squirrel came up my cellar stairs the other night (having fallen, I suppose, down my chimney) and crawled under the kitchen door. It put on a fine demonstration of its running, jumping, and gliding talents, before being directed to the oak tree in my front yard.

A census of northern Ohio mammals in 1600, or even in 1800, would have included, in addition to all the present natives, at least fourteen more, including three herbivores larger than the whitetail deer (elk, moose, and bison) and four large carnivores (gray and red wolves, mountain lion, and black bear).

There are slightly more than 200 bird species in northern Ohio, if we count not only those that live here during the summer and/or winter, but also those that pass through on their migrations north and south. Three of those species have been introduced from Europe (English house sparrow, starling, and pigeon), one from Asia (ring-necked pheasant), and one from western North America (house finch); the others are all natives of eastern North America (and for some of the migrants, of South America too). About thirty-one of these species (including all the aliens) are with us the entire year in northern Ohio. Some of my favorite year-round natives are the cardinal, blue jay, chickadee, goldfinch, nuthatch, downy woodpecker, and Carolina wren—all of which are frequently around my yard. Birds that build their nests around my home in the summer, but usually go south for the winter, include the robin, phoebe, Baltimore oriole, ruby-throated hummingbird, and flicker. Some northern-nesting birds that come south to Ohio to spend the winter are the junco, tree sparrow, white-throated sparrow, and redpoll. Migrants passing through here in their north–south travels, but not stopping for long, include many warblers, several sandpipers, and the yellow-bellied sapsucker.

Fish species of the Lake Erie watershed in northern Ohio number slightly over 100, out of the 166 species found in the entire state. Fish are found, like other animals and plants, only in habitats that are suitable in the present and accessible in the past. History is to be read in the distribution of fish in the two watersheds of Ohio, especially where their streams approach one another in northern Ohio. The common darter in the tiny streams flowing into Lake Erie is the Johnny darter, which is also found in the tributaries of the Ohio River, and hence throughout the state. The history of the Johnny darter is such that it has been able to bridge the present watersheds. But two other darters, the banded and the variegated, reach into northern Ohio only in the tributaries of the Ohio River. They are only a few miles from the rivers flowing into Lake Erie, but searchers have not found them there. As I watch the Johnny darters in Chance Creek, I think how time has brought them to their home—much as the flossy cottonwood seeds drift on the breeze, waiting to be planted on some future gentle shore.

When the Welland Canal was completed in 1829, the planners had ships in mind—to join Lake Erie and the upper Great Lakes with Lake Ontario and the St. Lawrence route to the Atlantic. But certain fish did not know that they were supposed to pay a fee for use of the locks. Niagara Falls had been an impediment to them as well as to ships. Three Atlantic species have wandered into Lake Erie since the nineteenth century—the sea lamprey, the alewife, and the freshwater eel. The eel is migratory between freshwater and salt water. Reversing the ways of the salmon, it breeds in the ocean but lives most of its adult life in freshwater. A few eels seem to be negotiating the canal and the entire St. Lawrence, keeping their species going in small numbers in northern Ohio.

Two Eurasian fish occupy the waters of Lake Erie and its tributaries. Goldfish escaped from captivity in the late nineteenth century, and carp, with which they hybridize, have been widely introduced.

About twenty-seven amphibians (eleven toads and frogs, and sixteen salamanders) inhabit the Oberlin region, and sixteen reptiles (seven turtles, one lizard, and eight snakes). All these are natives. The invertebrates are too numerous for anyone yet to count. Most are natives, but there have been important Eurasian introductions, intended or otherwise. The honeybee reached northern Ohio before the New England settlers, having flown here from the Atlantic colonies, where it had been introduced from Europe. Houseflies, bark beetles, gypsy moths, and so forth have come by various means.

Every animal has a story to tell that connects our time and present place with the entire history of the world. I went to the woods of Kipton Creek on a cold first day in February, to see how green the mosses are when all the world is gray. The snows had melted, and the ground was rich in Catharinea and haircap moss, white and windblown moss, and even sphagnum by a little pool. But another presence was in the woods that day, one that had a strange tale to tell of dark continents drifting in far distant time. On a tree limb overhanging the stream crouched a lone opossum, looking quite miserable in the damp cold, with a dripping nose, and with tail, feet, and ears largely bare of fur, as though it had not expected such weather.

The opossum is the only marsupial native to most of North America today. Its babies are born prematurely, the size of bees, after a gestation of only thirteen days, and are nurtured for the next three months, like baby kangaroos, in a pouch containing nipples. Marsupial mammals were once prevalent on every continent, but with the evolution of placental mammals, they were eliminated from much of the world. The ancestors of the opossum, however, were aided by extraordinary geologic events.

The present fauna of northern Ohio is the result of 500 million years of animal evolution since Cambrian times. Nineteenth-century geologists noted, in the geologic column of rock strata, two regions where abrupt changes occurred in the fossil patterns—where fossils previously plentiful abruptly disappeared. One of these was between the Permian and the Triassic rocks, where trilobites and so many invertebrate species disappeared; the other was between the Cretaceous and the Tertiary rocks, where many reptiles and all the dinosaurs vanished. These discontinuities permitted a division of the earth's history into three major eras: the Paleozoic (Age of Invertebrates), the Mesozoic (Age of Reptiles), and the Cenozoic (Age of Mammals). Many geologists today believe that the Permian–Triassic and Cretaceous–Tertiary extinctions were the result of major catastrophes in the earth's history (large volcanic eruptions, and a collision with a meteor).

Prior to the first of these catastrophes, all animals and plants probably shared a single supercontinent, Pangaea, which spanned an enormous range of latitudes and held land masses at latitudes very different from those they occupy today. At Salto, Brazil, within the present Tropic of Capricorn, an unusual block of Precambrian granite was discovered in 1946, having a shape identical to that of numerous granitic outcrops and islands in the recently glaciated lake country of Canada and Minnesota. In North America such rocks, with a gentle, smooth incline on the front side and a steep and sometimes ragged slope on the back, indicate the direction that continen-

149

tal ice flowed over them. Another such rock was found in Africa, matching in direction the one in Brazil. These rocks suggest that Africa and Brazil were joined in Carboniferous and Permian times, and were scoured by a common continental glacier. They also indicate that regions now tropical had antarctic climates then. What are now the northern regions of Eurasia and North America did not see continental ice until 270 million years later in the Pleistocene, and must have occupied warmer climates in Pangaea. At the end of the Paleozoic, however, Pangaea began to pull apart.

The breakup of Pangaea proceeded throughout the Mesozoic era, but substantial connections remained among all the present continents, so that reptilian fossils, for example, can be found in Antarctica as well as in all the other continents of the world. During the Cenozoic, however, the land masses of South America, Australia, and Antarctica separated entirely from the other continents, and the evolution of mammals proceeded in different ways on different continents.

Australia, carrying a diverse marsupial fauna, separated from the other continents before the evolution of placental mammals; seven families of marsupials survived in Australia, and that continent saw no placental mammals until bats, rodents, people, and dogs arrived in the Pleistocene. South America pulled away from Africa a little later, after placental mammals had started to evolve, and it was able to sustain, in its isolation, two families of marsupials as well as a unique evolution of placental mammals.

The ancestors of our opossum were riding the South American boat when, in the late Cenozoic (Tertiary), it ran up against Central and North America. Some of the opossums jumped ship and headed north, along with placental sloths, strange glyptodonts (which might be described as huge mammalian turtles), armadillos, capybaras (large rodents looking like little pigs), and porcupines. Fortunately, no immigration officers were at hand to regulate this motley crew, though competitors in the north and other facets of natural selection eventually took their toll. Not to be outdone in the search for better lands, northern animals headed south: The llamas of the Andes came from the Rockies (while their brothers, and all their camel cousins, became extinct in North America); tapirs (which also went extinct in the North, but survived in their new southern world); raccoons; deer (close cousins of the present whitetail deer); wolves; bears; and American lions, including the mountain lion (cougar), which today ranges through the mountains from British Columbia to Patagonia—and until the 1830s inhabited Ohio as well.

Of the South American adventurers, only the opossum survives in northern Ohio today. The porcupine (a creature mostly of northern or mountain conifer forests) was present in small numbers until the woods were cut down, for its bones have been found among Native remains at the Franks Site above the Vermilion River. The armadillo survives in the southernmost part of North America. It seems a little odd that it is the marsupial that has managed to adapt to the long journey north and to the great changes brought about recently by people, when in earlier history all the marsupials of Eurasia and North America gave way to placental competitors. The opossum is not well adapted to a cold climate. Its tail, feet, and ears can easily suffer frostbite in freezing weather.

Some think that the opossum has only recently arrived in northern Ohio and that it is still extending its range northward, but I believe that it has been here for hundreds of years, and is not going anywhere unless there is a change in climate. The explorer La Salle, traveling across the northernmost part of

Indiana in December 1679, killed two opossums, and it seems likely that he could have done so just as easily in northern Ohio. The Native Americans living along the Vermilion River left bones of the opossum at the Franks Site. Recent observations suggest that opossum populations decline when winters are severe, as in 1976 to 1978, and increase again with more moderate winters.

On a Sunday morning last November, three whitetail deer came into the little catalpa woods behind my house. They often travel between the large woods to the north of town and the waters of Plum Creek to the south—but especially on Sundays, when the traffic they must cross is light. A few days before, I had raked the catalpa, oak, and maple leaves from my lawn, and carried them into the woods to give their nutrients back to the trees from which they had come. The deer browsed on the few remaining green leaves on the trees and on herbs at the edge of my garden, and then lay down, with legs folded beneath them, on the soft piles of leaves from the lawn. They rested for half an hour on the leaves, their eyes surveying my kitchen windows, and ears alert for any barking from my little dog. Then they were gone, back toward the safer woods to the north.

The whitetail deer is the most ancient of all the native mammals of northern Ohio. It has lived here for 3 million years, except when continental glaciers moved it temporarily south—or when overzealous hunters, in the late nineteenth century, nearly drove it from the state. It has survived, in its present form, extraordinary changes in the mammalian life of North America. It has shared the leaves, grass, and fruits of the land not only with moose, elk, and caribou, but with camels and horses, mastodons and mammoths, and giant bison and giant sloths—while evading the attacks of saber-toothed tigers, chee-tahs and hyenas, and various wolves, bears, and lions.

Except for the opossum, all the present mammals of north-ern Ohio have descended, like the whitetail deer, from an evo-lution in the northern world of Eurasia and North America. Until the "recent" postglacial extinctions, the Ohio fauna was more cosmopolitan, with mammoths and mastodons of African origin, and sloths from South America. Postglacial fossils of those and many other mammals have been found in northern Ohio, but the glaciations removed whatever mammalian fossils were left prior to that time. The record of the whitetail deer and of other mammals of northern Ohio, during the 3 million years prior to the Wisconsin glacier, has to be reconstructed from sites in the western states and Florida, and from a few unglaciated sites in the northeastern states.

During a period from 3.5 to 1.9 million years ago (called the Blancan age from a site on Mount Blanco in Texas), North America shared with Eurasia a diverse mammalian fauna. The beginning of the Blancan age is marked by the appearance of modern (one-toed) zebras and horses. Fossils of evolving horses, going back to the small "dawn horses" of 55 million years ago, have been found in both North America and Eurasia, indicat-ing that those two continents remained connected throughout most of the Tertiary period—and that horses have always been good travelers! The fossil horse lineage is most complete in North America, so the center of horse evolution may have been here. The zebras of modern Africa could well be descended from American Blancan zebras. Pleistocene horse fossils have been found in Ohio, but all American horses went extinct in postglacial times.

The horses and whitetail deer shared the Blancan landscape

with mastodons and camels. The mastodon evolved in Africa and reached North America about 14 million years ago by way of Eurasia. But early camels and llamas lived only in North America, and not until the Blancan age did they begin to migrate to Eurasia, Africa, and South America. A great many remains of Pleistocene mastodons have been found in Ohio, including three near Oberlin, but no one seems yet to have found any remains of camels in this state. In postglacial times, the mastodon went extinct throughout the world, and North America lost all its camels as well.

Other Blancan mammals included the modern North American beaver, bobcat, and coyote; the porcupine, which had arrived from South America; cheetahs and hyenas like those of Africa today; and small pandas like those of the Himalayas. There were also dirk-toothed tigers, with long fangs in the upper jaw, and a wolf with massive jaws for crushing bones—the "bone-eating dog." In those days, deer resting in the woods had far more than a little dog to watch out for!

About 1.9 million years ago, the mammoth wandered into North America from Eurasia, to join the mastodon, which had come 12 million years earlier. The presence of the mammoth in North America signals the beginning of the Irvingtonian age (after a site in Irvington, California), which lasted until 0.4 million years ago and witnessed early Pleistocene glaciations.

Irvingtonian times saw the evolution of the modern muskrat, vole, lemming, and hare as well as the migration of the black bear and wolverine from Eurasia, along with the shrub ox and Soergel's ox—the first cattle (Bovidae) to reach North America. The bones of many of these Irvingtonian mammals have been found in the Port Kennedy Cave in Montgomery County, Pennsylvania.

For nearly 2 million years, both mastodons and mammoths

lived in North America. In Eurasia, where still other elephants inhabited the warmer regions, mammoths tended to
be restricted to the arctic environment, but in North America
they were able to invade the warmer grassy plains as well.
The flat molar teeth of the mammoth, with many transverse
ridges, became increasingly well adapted to grinding grasses,
while the sharper teeth of the mastodon were better for sever-
ing woody twigs. These two giant herbivores did not compete
very much for habitat, the mammoth living mostly in grassy
regions and the mastodon in forests (fossils of black spruce,
hemlock, and larch have been found in mastodon rib cages).
While mastodons seem to have been much more common
than mammoths in postglacial northern Ohio, remains of
mammoths have been found in Huron and Cuyahoga counties. Beautiful reconstructions of mastodon and mammoth
skeletons are on display, side by side, at the Cleveland Museum
of Natural History.

About 400,000 years ago, the Eurasian bison was able to
find a route from Alaska past the ice of the Illinoisan Glaciation
into the open country of North America. A great number of
northern Eurasian mammals accompanied the bison: the
brown bear (ancestor also of the grizzly bear and Kodiac bear);
the wolf, red fox, and arctic fox; the ermine, snowshoe hare,
and mountain pika; the mountain goat and bighorn sheep
(joining the American antelope, or pronghorn, which had been
here since Blancan times); and the elk, caribou, and muskox
(which inhabited woodland as well as tundra). The last of the
large Eurasian mammals to find its way past the ice to the Great
Lakes region was probably the moose, which may not have
arrived until the Wisconsin ice had retreated far to the north.

The coming of the bison marks the beginning of the Rancholabrean age in North America (named for the Rancho La

Brea tar pits in Los Angeles). The tar pits trapped more than 1,000 saber-toothed tigers and 1,600 dire wolves, along with 40 other species of mammals and 120 species of birds. The saber-toothed (and scimitar-toothed) tigers ranged throughout the North American West and all the way to Brazil and Argentina; close relatives inhabited Eurasia as well. They hunted in open grasslands, but also made use of caves; baby mammoths were one of their prey. They occurred as far northeast as Tennessee, but as far as we know, did not reach Ohio.

The remains of many other Rancholabrean age animals, however, have been found in Ohio. The bison, which ranged through woodland as well as grassland, left its bones here, as did the moose and muskox. Ancient elk and caribou bones have been found in Kentucky, just south of Cincinnati. The grizzly bear left remains all around us, in Kentucky, Ontario, and the Appalachians, so it must have lived in northern Ohio as well. Other Ohio fossils include the tapir and two kinds of pig-like peccaries, whose ranges tended toward the south, and a giant beaver and giant sloth, now both extinct.

The giant sloth was descended, like the opossum and the porcupine, from South American ancestors. It was certainly the most improbable of all mammals to share the woods and fields of Ohio with the swift whitetail deer. The size of an ox, the giant sloth had found its slow way from South America to Alaska along the Pacific coast and up the Atlantic coast to the northeastern states. The first remains of this animal to be discovered were its huge claws, and Thomas Jefferson, amateur paleontologist as well as Greek scholar, gave the animal the generic name *Megalonyx* (Greek for "great claw"); it is now known to science as *Megalonyx jeffersonii*, or the Jefferson ground sloth. *Megalonyx* remains have been found at Norwalk, Ohio, and in a swamp in Holmes County. The Holmes County skeleton lay on top of marl deposited by the Wisconsin glacier, in peat that must have accumulated after the glaciation—indicating that the sloth had come into our region sometime after the retreat of the Wisconsin ice. Very shortly thereafter, the Jefferson ground sloth went extinct.

So did about fifty other large mammals. The Rancholabrean age, a golden age for North American mammals, came to an end. Not since the Cretaceous–Tertiary extinction of the dinosaurs did the world see such a loss of magnificent animals, and for North America the toll was even worse, for many animals that survived elsewhere were extinguished here. Gone were all mammoths and mastodons, giant sloths and giant beavers, saber-toothed tigers and dire wolves; gone were our camels, our horses, our tapirs. One new species (our own) had recently arrived from Eurasia, to experience briefly this passing fauna. The Wisconsin glaciation was over, and rapid changes were occurring in the climate, in the landscape of ice, lakes, and rivers, and in the vegetation. We will never know to what extent it was people or natural environmental changes that caused this faunal catastrophe.

Our present fauna connects us with all time, and with the whole world as well. As I walked a footpath in southern England, a startled bird flew out of the hedge beside me, crying an alarm that sounded to me like the call of the American robin. It was an English blackbird. In spite of the differences in the colors of their feathers, the English blackbird and the American robin are close cousins, both members of the genus *Turdus*, and their alarm calls are very similar. Biologists can argue endlessly about whether a population of animals from one area is the same species or not as that from another region, but the red fox that runs through the fields of Ohio is essentially the same red fox that crossed my back garden one day in Oxford. The same animal has adapted itself to life in rural Ohio and urban England. The gray wolf, beaver, lynx, wolverine, and stoat (ermine) of America are essentially the same as those of Eurasia. The American elk is almost the same as the red deer of the highlands of Scotland. They are all connected to one another through the geography of time.

Some of our local animals have global lives. I always find it hard to believe that the loon of the summer Canadian lakes spends its winter on the Atlantic Ocean, and that such birds as the bobolink, red-eyed vireo, and many of our warblers have South American citizenship. The tiny Blackburnian warbler that collapsed one spring in my garden had flown all the way from Costa Rica or Peru; a solitary sandpiper, expired on the banks of Elk Creek in early May, was on its way to Canada, perhaps from Argentina. When animals have such global lives, they become especially vulnerable to environmental changes, because they rely on so many different ecosystems.

Birds migrate to find nesting sites in summer and food in winter. Those that leave our northern land for the winter are not so much avoiding cold (they are supremely well insulated), as seeking adequate food supplies. As insects disappear for the winter, insect-eating birds must leave for other lands, unless they are nuthatches or woodpeckers adapted to finding dormant insects in the bark or wood. The short days of winter greatly reduce the time available for feeding, and the long nights require energy for heat for extended periods without food. Sometimes the snow or ice of winter makes it impossible for a bird to feed, even if its food is still available. Most birds cannot find or reach the seeds or insects that lie beneath the snow. A loon or heron cannot feed when the fish are beneath the ice, and the poor loon is unable to become airborne unless it has sufficient clear water from which to take off. The turkey vultures of northern Ohio have to move south when the snow cover eliminates thermal up-drafts from the heated land, on which they soar in looking for food in other seasons.

The energy cost of migration is greater, proportionately, for smaller animals than for bigger ones, so the feats of small birds (such as vireos and warblers) in flying to South or Central America seem all the more remarkable. The prize for smallness combined with long distance, however, should probably go not to a bird, but to a butterfly.

The monarch butterflies often cross Lake Erie from Canada in the last days of September. On two occasions (September 27, 1970, and September 29, 1972) I was walking near the Lake Erie shore at Oak Point, through fields full of white asters, white snakeroot, goldenrod, and purple New England asters, when monarchs appeared everywhere about me, some feeding on the New England asters, others lying exhausted on the ground. Along the roads, many were killed by speeding cars. In those days, we knew only that the monarchs flew long distances toward the south and west. Fred Urquhart in Toronto had been studying them since 1935. He had developed a technique for banding the fragile wings of the butterfly, and with the help of correspondents who recorded recoveries in different parts of North America, he showed that the lines of flight of most monarchs pointed toward Mexico. In January 1976, he found the monarchs' wintering site in the Neovolcanic Mountains of Mexico, at latitude 19 to 20 degrees north, and at elevations of 8,800 to 16,400 feet.

Animals are full of surprises. We thought that the monarchs flew south to avoid the northern winters, but if 300 feet of elevation up a mountain is equivalent to 1 degree of latitude toward the pole, the monarchs even at 8,800 feet in Mexico are considerably farther "north" than they were in northern Ohio.

There has to be some other advantage in Mexico that northern Ohio or Canada cannot offer. But what do the monarchs know of this? The ones that fly to Mexico were born somewhere farther north; it was their deceased parents or grandparents who last were in Mexico.

Most adult monarchs survive the winter on trees in the Mexican mountains; a much smaller number overwinter at Pacific Grove, California, or in Florida. In spring they mate and begin the migration north. During the trip north, a female lays its eggs, one at a time, on milkweed leaves, and if it is lucky it may make it all the way to northern Ohio or Canada, usually with tattered wings. It will die, however, before the summer ends. Later arrivals into the north come with fresher wings (and no direct memory of Mexico), for they are the new generation emergent from the eggs laid along the migration route. It takes four to six weeks altogether for the eggs to hatch and for the caterpillars to feed on the milkweed leaves, make their pupal cases, and metamorphose into adult butterflies. If conditions are right, large local populations of monarchs can develop in northern Ohio during the summer, before the fall migrations come through from the north. The size of the migrations themselves vary from year to year. The summer of 1892 must have been a good one for monarchs, for such a large number came across Lake Erie to Cleveland on September 19, 1892, that the Cleveland *Plain Dealer* featured an article about it the following day, stating that some citizens were afraid that the butterflies might be cholera germs in disguise.

Some of the monarchs from Ohio cross the Appalachians from north to south (and hence also from west to east) to reach the coastal states before turning west toward Mexico. Monarch migrations have been reported through Newfound Gap and Indian Gap in the Great Smoky Mountains. In late October,

I once found near the summit of Mount Collins two monarchs that missed the gaps by 2 or 3 miles and, traveling a bit too late in the season, did not make it over the higher reaches of the Smokies; curiously, each had expired on a soft, rich bed of *Hypnum crista-castrensis*, the most luxuriant moss in the mountains.

In the northern winter that the monarchs and warblers flee, the land holds time silently in the long shadows across the snow. Deep serenity fills the winter woods, with intimations of timeless dimensions unknown at other seasons. The energies of life move quietly in this domain, sometimes leaving tracks like soft fossils in the snow. One night a rabbit traversed a woods from east to west, and a fox from north to south; their tracks crossed in space but not in time, and remained in calm independence in the morning light. Perhaps Orion saw each pass as he hunted across the sky.

The winter is the quiet season because so much of the plant and animal world is dormant. I enjoy the calm of such times, when no mosquitoes or flies buzz around my face, and poison ivy has only winter stems devoid of much authority. The Christmas fern, hemlock, marginal wood fern, and polypody remain green all winter, as do a few herbs like waterleaf and wild stonecrop. But most of the trees and herbs have lost their photosynthetic leaves, and those that retain them are able to produce very little food in winter. With plant growth suspended, many of the animals choose to become dormant themselves. Those that stay active must feed on the produce of the previous summer—seeds or fruits, dried stems or roots—or on one another.

The adaptations of animals to winter are almost beyond belief. I paused near a woodsy pool in March to listen to the duck-like quacking of a wood frog hidden in the brush. Two weeks before, that vibrant frog was a silent, frozen hulk, buried beneath the snow and leaves and mud. Like many other cold-blooded animals, the wood frog is able to produce antifreeze (in this case, glucose) to keep its living cells from freezing, while allowing ice to form in the blood and body fluids around the cells. The frozen frog appears to be one mass of ice, but all the ice is around its body cells, and none is actually in them. The spring peeper, gray treefrog, and garter snake all perform a similar trick to keep their cells from being disrupted by ice crystals. They are like the hardy northern trees that freeze, but only in their extracellular spaces.

Many insects, too, survive the winter by restricting freezing to the extracellular fluids of their bodies. Others avoid freezing altogether with glycerol or ethylene glycol or with special proteins that allow their cells and body fluids to resist ice formation when "super-cooled." Different insects spend the winter at different stages in their life cycles: as eggs (for example, tent caterpillar, grasshopper), larva (woolly bear, goldenrod gall fly), pupa (Promethea moth), or adult (mourning cloak butterfly, ants, bees, wasps).

Other cold-blooded animals avoid the possibility of freezing by retreating deep into burrows below the frostline (earthworms, American toad), or by staying in ponds of water (fish, green frog, leopard frog, bullfrog, many turtles). Many snakes form large colonies in a winter den, producing collectively enough heat to prevent freezing; honeybees do the same.

Some warm-blooded animals reduce their heat losses and energy needs in winter by lowering their body temperatures and becoming inactive. The northern Ohio mammals that do this for extended periods include the bats, chipmunk, raccoon, and woodchuck. In all cases, enough heat is generated by the hibernating animal to keep its body temperature above freezing. The woodchuck's body temperature drops to about 12 degrees centigrade, and its heat production to 5 percent of normal; its heart rate is four to five beats a minute, while it breathes just once every one to five minutes. Chickadees conserve energy by allowing their body temperatures to drop to 30 degrees centigrade for short periods. Most birds resort to shivering to keep their body temperatures from falling, while mammals have brown fat cells to generate extra heat when needed. White-footed mice become more social in the winter, sharing the heat of one anothers' company, and spend much of their time beneath the insulation of the snow.

With the first melt of the snows in March, the silence of winter is broken by the voices of spring peepers, first a lone male in the evening dark, and then great numbers competing with one another to own a segment of the shallow pool in the brushy field east of my home. It always seems as though the earth itself has come alive by a spontaneous generation from its muds. The calls of the frogs are echoes, across 200 million years, of the earliest vertebrate voices heard on the land. The spring peepers emerging from the mud are continuing a ritual that goes back to ancient amphibians first crawling from the waters. I have heard the first spring peepers as early as March 6 (two weeks before the vernal equinox) or as late as March 29. Sometimes the first ones are silenced for a week or two by a return of cold weather. A week or two after the first peepers, the gray treefrogs begin their raucous calls, and later still, the ethereal trill of the American toad rises to the stars—perhaps the most beautiful of all the sounds of night.

Like birds migrating between North and South America, the frog and tadpole are moving between two worlds, trusting the survival of their kind to the stability of two different ecosystems. The adult frog and juvenile tadpole have evolved separately in those worlds while joined in one life cycle. How has such a thing come to be? The tadpole and frog are attuned to different food sources, the algae of the water and the insects of the land, and on average each enjoys some sort of advantage in doing what it does. Ecologists point out, however, that in linking aquatic and land communities more completely together, amphibians may be helping both, by harvesting nutrients that have leached from the land and recycling them back to the land again. Does an animal evolve to aid its world? In helping to stabilize their environments, the frog and tadpole are conserving worlds that are conserving them. In some extended sense, the life of every organism is both the result and the cause of the world in which it lives.

Two great themes run through the ecology of the living world: the flow of energy and the recycling of materials. The energy of life is captured by the chlorophyll of plants—the green film that intercepts the sunlight and awakens the earth from a moon-like silence into the humming busyness of a summer's day. The honeybees bustling about my garden, the darters dashing in the little pools of Chance Creek, and the warblers flitting from branch to branch through the woods at Kipton Creek—all have their energy from the sun, passed onto them by channeled electrons in the chloroplasts of plants.

The sunlit plants could grow without the animals, but not the animals without the plants. Plants have coevolved with animals and learned to depend on them for pollination and seed dispersal. A woodland flower may coat its seeds with nutritious "elaiosomes" to induce ants to carry them to their nests—distributing future flowers around the woods. But the elaborate ornamentations of ecology do not alter the primary fact that the energy flux of the forest originates in the green leaves of its plants.

As the energy flows, materials must be recycled. Growth alone would soon cease for lack of essential nutrients, were plants unwilling to give their substance back to the earth and air. The air we breathe contains only about .03 percent carbon dioxide, an amount so small we do not notice it, but the photosynthesis of plants is utterly dependent on it. Green plants respire and return a certain amount of carbon dioxide to the air from the sugars that they have produced, but for the recycling of much of their materials, plants are dependent on fungi and microorganisms. The mushrooms that spring up after a rain from the fallen leaves and logs on the forest floor—the red chanterelles and deadly amanitas—are the fruiting bodies of hidden masses of fungi digesting and recycling the plant material. Animals speed the recycling by foraging on plants and on the fungi and microorganisms. A mushroom is often infested with beetles and flies that are consuming it, and in the soil, nematodes and earthworms work on the fungi, microbes, and plant remains. In principle, however, the other kingdoms of life could manage the recycling by themselves. Neither the ecological flow of energy nor the recycling of materials requires the presence of animals.

The biological character of a land, therefore, starts with the plants rather than with the animals. The evolution of land animals in the Devonian period was dependent on the establishment of a terrestrial flora. When E. Lucy Braun classified the ecology of eastern North America, she studied the distribution of trees, not mammals. In a woodland, the canopy trees are the first to intercept the sunlight, and they establish the character of the land and life below. In Braun's classification, northern Ohio is (or was) part of the beech–maple forest. No one has ever tried to call it a "mouse–squirrel–wolf region," because the specific animals have a less general effect than the canopy trees on the overall ecology of the land. Nevertheless, a red squirrel carries pine cones around, ants distribute the seeds of the spring flowers, and deer browse the lower limbs of trees. Every animal has an effect on the land.

Thoreau knew all this, from his walks around Concord. But the first in North America to record how land, plants, and animals affect one another through time was a young geologist–biologist named Henry Chandler Cowles. As a geology student at Oberlin, Cowles had learned how landforms change spontaneously with time. He brought this dynamic viewpoint to his graduate studies in botany at the University of Chicago in 1893. He reasoned that if landforms are changing by natural processes, so must the vegetation of the land, as the latter depends on the former. He came to see also that the relations between land and flora are interactive: that as changes in the land affect the vegetation, so too the vegetation affects the way in which the land changes.

As a place to study the vegetation of a dynamic landscape, Cowles chose the sand dunes at the southern end of Lake Michigan. Because the surface level of Lake Michigan has fallen somewhat in postglacial times, the dunes farther from the present lake shore have had a longer time to develop than those closer to the shore. Changes in the dune structure with distance from the shore could therefore be viewed as changes over time—so that one locale could display all at once the landform changes that have happened on a lake shore over a long time.

Cowles saw that the vegetation changed with distance from the lake shore, from no anchored plants at all at the water's edge, to beech–maple woods inland. The lower beach, which was washed by summer waves, had no plants. A middle beach, which was free of waves in summer but washed by higher water in winter, had (in summer) a few rooted plants; these were annuals (which could grow from seed and produce new seed in one summer season), and they were succulents (which could withstand the dry conditions of the summer beach). An upper beach, which was free of waves in both summer and winter, allowed the growth of biennials and perennials—plants that live more than one summer season. A succession of stages of vegetation could thus be seen in moving from the water's edge toward the first dunes.

With the formation of dunes, the vegetation changed still further. But this was not a simple one-way relation. Cowles saw that the geologic process of forming dunes by wind-driven sand was in fact vegetation-dependent: The formation of dunes required the presence of perennial plants to catch and hold the blowing sand, and the ultimate shape, area, and height of the dunes depended on the specific plants involved. Dunes formed by *Elymus* grass (wild rye) were small in area, as this grass does not spread rapidly by rhizomes, while those formed by *Ammophila* and *Agropyron* grasses, which do spread readily via rhizomes, were more extensive. Once dunes were stabilized by grasses or other herbs, woody plants could begin to grow, and because their roots could grow deeper, the dunes could become higher. Hence in succession from the lake shore (and in age), grass dunes would be replaced by dunes with willows, then those with cottonwoods or pines (adapted to dry conditions), and finally, as soil developed and moisture increased, by woods of beech and maple trees.

Cowles's study, published in 1899, became one of the foundations of the new science of ecology. Having shown that plants change with the age of a dune landscape, Cowles was confident that the animals must change as well, because they are dependent on the plants. In 1913, Cowles's student Victor Shelford recorded the way that the Indiana dunes showed succession of animals as well as plants. Shelford found, for example, that grasshoppers, digger wasps, and sand spiders are prevalent in the beach grass and cottonwood dunes close to the lake, but are not found at all in the wooded areas farther inland; while sowbugs, earthworms, and wood snails are prevalent in the beech–maple woods, but are absent from the lake-front dunes. The ants and tiger beetles found near the lake are different species from the ants and tiger beetles found in the woods.

The Cowles–Shelford studies of our Great Lakes dunes serve as a model of how the land, with its plants and animals, will tend to change with time. Geologic processes and the lives of flora and fauna are all linked together in an interactive and evolving community of nature. Whenever a new factor is introduced, whether by natural occurrence or by human intervention, we can expect many animals and plants to be affected, but it is impossible to predict exactly how the changes will unfold.

The cutting of the forests of northern Ohio led to an invasion of Eurasian plants into the newly opened areas. The same did not happen with animals. Domestic cattle, horses, and pigs, for example, have not gone wild and taken over the fauna of our region—though feral pigs (wild boars) have had a considerable effect in the southern Appalachians. The only Eurasian mammals that have become established in northern Ohio are the house mouse and the Norwegian rat, and they tend to be restricted to the environs of human habitations. I have no European squirrels or rabbits running around my yard. They are all the native red and gray squirrels and cottontail rabbits. The mice that succeed in getting into my house are almost all native white-footed mice instead of the gray European mice. The reptiles, amphibia, and fish of northern Ohio are almost all native species. Three European birds (house sparrow, starling, and pigeon) and many Eurasian insects, however, have greatly affected the American land.

The extinction of large native mammals from northern Ohio in modern times has resulted not from competition by European counterparts, but from the destruction of habitat by increased human populations and from a war on wild animals. The war began in Europe and was carried to North America in the seventeenth century. In Roman times, Britain had bears and wolves, wild boar, and lynx and beaver, all of which had disappeared by the eighteenth century. The beaver was hunted to extinction for its fur, while the wolf and bear were driven out by the farmer. I was talking to a friend one evening outside her farmhouse in northern Maine, when from her upper field near the forest the distressed bellowing of a mother cow shattered the night air. We rushed with flashlights to the scene, but were too late. A black bear had taken the mother's calf and had disappeared into the darkness of the forest. No farmer can enjoy such acts of unfettered nature. When there are few axemen and farmers to interfere, the forest preserves its own, but when there are too many people, the bear has no chance.

A few centuries ago in Ohio, the Native American farmer had no domestic livestock and valued the bear highly for its meat, along with the wolf, mountain lion, and large herbivores: whitetail deer, elk, and bison. The bear was plentiful, and especially prized. But as soon as white farmers moved into Ohio from Virginia and New England, the war on the black bear began, and on the wolf, and even on the whitetail deer and the gray squirrel. Ohio had been a state for only four years when, in 1807, its legislature required the citizens to kill gray squirrels or pay a tax in lieu of scalps. Farmers organized large-scale, coordinated hunts to exterminate wildlife. A Franklin County hunt in 1822 claimed to have killed 20,000 gray squirrels. In 1818, a mass hunt near Hinckley killed, in a single day (Christmas Eve!), 21 black bears, 17 wolves, and 300 whitetail deer. The Lorain County commissioners offered a bounty for wolves in 1827. John Shipherd still heard wolves near Elyria in 1831, but not for long. The bear and the mountain lion had probably already disappeared, along with the bison and the elk. Soon the wolverine, lynx, bobcat, otter, and beaver were also gone.

Wildlife vanishes from a region not only when it is hunted to excess, but when its life habits and habitat—its ways of feeding and protecting and reproducing itself—are too much disrupted. A bear cannot fatten itself, for winter hibernation, in a field of summer wheat!—though a raccoon, in a field of corn, can do so very nicely. When the forests are cut down for agricultural fields, the wolverine and porcupine, the lynx and bobcat will disappear, whether or not people take any further interest in them. When wetlands are drained, the beaver, otter, muskrat, and mink are without a home. Humanity affects the complex lives of animals in ways we can easily overlook or may not yet be able to understand. It is no mystery why wolves are absent today in the small woodland areas of northern Ohio. But it is not at all clear why the wolf disappeared about 1925 from the rugged wilderness of the Great Smoky Mountains or why its smaller cousin, the coyote, is today spreading back to areas long vacated.

The past two centuries have altered also the nonmammalian fauna of northern Ohio. Woodland species have been greatly reduced or eliminated. The most famous case is that of the passenger pigeon, which was once our most abundant bird. The passenger pigeon disappeared from the Ohio countryside in 1900; the last one died in the Cincinnati Zoo in 1914. When I see bright fruits hanging all winter, untouched, on our native shrubs today, I wonder if they were once eaten by the passenger pigeon. The elimination of one species has consequences for many others.

The clearing of the forests, and the plowing of fields for crops, increased greatly the amount of rainwater that ran rapidly off the surface of the clay soils and into the streams, instead of seeping vertically through the soil. All the streams today are muddy with clay silt, even in dry seasons, and it is hard to believe that until 1850 they were clear, even in times of flood. The consequences for nearly all aquatic species of plants and animals must have been profound. Excessive commercial fishing in Lake Erie reduced the sturgeon and lake trout almost to extinction long before the accumulation of nutrients from agricultural runoff and sewage threatened the whole life of the lake. The drainage of wetlands bordering the streams and lake and the pollution of waters from industry and from agricultural herbicides and insecticides have also had more effects than we yet can know.

In 1850, while the passenger pigeon was being extinguished by hunting and by the cutting of our woodlands, a few English house sparrows were brought to New York. They died. But the next year, a few more were brought, and they began to conquer the American town and countryside. The English sparrow (*Passer domesticus*) is one of fifteen species in the genus *Passer*, which probably originated, like humans, in tropical Africa and, like humans, inhabited open forests or savannas. Eight of the fifteen species live in close proximity to people in Eurasia and Africa, breeding often in the eaves of inhabited buildings. The "English" sparrow had, quite on its own, become comfortable with life throughout much of Eurasia before it was brought to New York. Soon it was introduced to other states, including Ohio, and to South America, Australia, New Zealand, and

South Africa. It has become a bird for all continents as well as all seasons.

In 1890, sixty European starlings were released in Central Park, in New York City. Today the starling ranges from the Atlantic to the Pacific across the United States and southern Canada. They are communal birds, feeding and roosting in flocks. Their night roosts in Oberlin are in new-growth woods with high densities of small trees, providing fine protection from the wind. When I bicycle in the evening down my street, they are often in great flocks in the trees above, forming a "pre-roost assemblage," calling cheerfully to one another before fly-ing off, in great waves of wings, to their night roost in the woods. They feed in short-grass meadows and fields on insects, seeds, and fruits, preferring the insects (with their high protein) during the spring when their young are being raised. The star-ling competes with native grackles, black birds, cowbirds, and orioles that feed in a somewhat similar manner; as a newcomer to a large continent, its competitive success has been remark-able. For nesting sites, the starling is partial to holes in trees, and may take over the sites of woodpeckers or wood ducks. Some think that the decline of the eastern bluebird is partially due to competition from starlings for nesting sites.

In 1975, a Christmas bird count was done at Buckeye Lake, Ohio; 14,092 birds were counted, comprising 86 species. Of these, 6,722 were starlings, and 2,321 were English sparrows, followed by pigeons, with 208. The pigeon was brought to the English colonies as a source of food in the seventeenth century (dovecots were popular on English farms), but because it is the Mediterranean rock dove in disguise, it has been happy to take over the stone (or wood) eaves of buildings in towns and cities. So in the Buckeye Lake count, three introduced birds from Eurasia outnumbered all the American birds by nearly two to one. Our American farms and towns, wherever they may be, are now—in their fauna and their flora—international communities, parts of the commonwealth and common history of the world.

As human history has greatly influenced the fauna of northern Ohio, so the fauna has influenced human history. Large mammals roamed North America for millions of years before there were any people. They did not need people to tell them where to go, or any human help in making their own trails. Some of these trails were probably as broad as roads. When people first found their way south of the glacial ice in North America, they would have followed trails previously made by herds of bison, caribou, and elk, as well as whatever trails the mammoths and mastodons made in those days. In flooded areas, short beaver paths and longer moose trails (once the moose reached our northern postglacial lakes) would have provided the way for future human portages between lakes, between the watersheds of streams, and around rapids in rivers. The animals find the way of least resistance between one place and another, and that is what people want as well. In the north woods, I have carried my canoe across hundreds of portages that probably began long ago as animal trails, and on many occasions have followed waterways or mushy paths made recently by the beaver. With their dams, beaver often make streams more navigable by canoe than they would otherwise be—though the opposite can sometimes seem to be the case as well! High in the Rocky Mountains, I have wondered whether I was following a human trail or an elk trail; often it was both.

So across North America today, many of our old roads, railroads, and historic routes—such as the Boundary Waters and other canoe routes in the United States and Canada—follow trails first made by animals. The trails became the pathways of

Native Americans and then the routes of European explorers, fur traders, soldiers, and settlers. In his *Historic Highways of America*, Archer Hulbert placed special importance on the trails made by the woodland bison through the Appalachian Mountains and between the watersheds of our rivers:

> It is interesting that he [the bison] found the strategic passageways through the mountains, so that the first explorers came into the West through gaps where were found broad buffalo roads; it is also interesting that the buffalo marked out the most practical portage paths between the heads of our rivers—paths that are closely followed today [1902], for instance, by the Pennsylvania and Baltimore and Ohio railways through the Alleghenies, the Chesapeake and Ohio through the Blue Ridge, the Cleveland Terminal and Valley railway between the Cuyahoga and Tuscarawas rivers and the Wabash railway between the Maumee and Wabash rivers.

From the perspective of an interstate-highway engineer, a trail through the woods may seem rather insignificant. But anyone who has tried to carry a canoe straight through a Canadian forest without a trail would agree that the difference between a path and no path can sometimes seem far more significant than that between a path and a superhighway. Animal trails are but one small example of how living nature has influenced our history, for our whole existence is dependent on the plants and animals of our land. But in leaving a permanent impression on the landscape, animal trails serve to remind us that we are of nature, and nature is an intimate part of us.

The history of the Great Lakes region, and indeed of the entire northern half of North America, was greatly affected by a single species of our native fauna, a rodent no less: the beaver.

The beaver was not even a uniquely American animal, for it ranged across northern Eurasia as well. But Europe had discovered a great commercial interest in its beaver shortly before it discovered America, and its own populations of this animal were becoming nearly extinct in the sixteenth century. When French and English fishermen began to work the coasts of North America, they were pleased to find that the Native people could supply them with beaver pelts, for which there was a high price in Europe.

The subsequent pursuit of the American beaver constitutes one of the world's all-time examples of how events of enormous consequence can unfold from utterly trivial beginnings. The beaver was wanted for no more essential purpose than to make fine hats for the European gentry. As an aquatic mammal, the beaver has an unusual inner fur beneath its longer outer hairs. The inner fur is made up of fine hairs with tiny barbs, which catch one another and provide an airy insulation for the skin, protecting it from the water. The inner fur is especially thick in the winter and in the northern populations of beaver. The trouble began when the Europeans discovered that the finely barbed inner fur could be made into felt, which, stiffened with shellac, became the material of the finest hats—not the finest hats for warding off the winter blizzard, but the finest hats of fashion.

There were large populations of beaver to be exploited in northern North America. The glaciations had created an ill-drained, flat topography full of lakes and streams for the beaver. The beaver's aquatic life, and sturdy houses with underwater entrances, protected it from excessive hunting by the Native people, as long as the implements used against it were made of stone and the demand for it was not great. But with the high European demand for the beaver came also the metal axes, guns, and traps for pursuing it relentlessly. The lakes and streams that were the beaver's home also provided canoe routes into the farthest reaches of its country. And the iron implements that the Europeans could offer the Native people in exchange for their beaver—the kettles, needles, knives, axes, traps, and guns—were of infinitely greater significance to them than a felt hat could be in Europe.

Thus the beaver provided the main impetus for the European penetration of the Great Lakes and the waterways of northern North America. Because the beaver populations of the East soon gave out, there was a great incentive to reach the larger populations in the West and North. The push began with the French coming up the St. Lawrence River in the sixteenth and seventeenth centuries; in 1660, Radisson and Groseilleurs explored as far as Lake Superior. A few years later, Hudson Bay was found to provide an alternative route to the interior of the continent. There was rivalry between French and English, between companies operating out of Hudson Bay and those using the Great Lakes, and, most important, between Native American people vying for a share of the commercial spoils. Soon the Ojibwa people were warring with the Sioux, pushing them out of the beaver's forest into the bison's plains, and the Iroquois were fighting with the Hurons and with the people of northern Ohio. Through the Iroquois wars and the transmission of European diseases, the beaver trade destroyed the northern Ohio people and changed the ecology of the land that the New England settlers would find.

The land—its rocks and waters, people, plants, and animals—are joined in a continually unfolding pageant through time. The scenes are changed by forces as tangible and immense as those that tore Pangaea asunder, or by energies as subtle and mysterious as the migrations of butterflies or the passions of human adventure. We participate in only a few moments of the pageant, yet each moment has the whole eternity within it. If we see the eternal, we will honor the moment and cherish the earth and all its wildernesses of life.

The creative forces of life and nature are everywhere in our world, from the mighty to the most sublime. Mountains are made, and in the mists of their forests, the lichens turn green and grow. The deer slip into my woods, rest on the leaves, and then are gone. Would I have any other wilderness than this? Which one would I have? The one the New England settlers knew? Or that of the Eries, or of the Hopewells, or of the Paleo-American hunters? Would it be the deciduous wilderness, or the earlier boreal one, or the ice of the glaciations, or the lands of hundreds of millions of years stretching back to the oceans of the Devonian fish? Whenever we live, we can sense the spirit of the whole, and revere the wilderness within.

As winter twilight settles on the center of the old town of Jundiaí in Brazil (where I am writing these last lines), I hear a familiar commotion in the trees of the church plaza—the high-pitched, frenzied chatter of hundreds of small birds settling down for the night. As a small boy in Ithaca, New York, I used to hear the same staccato chirping, by the same evening light (in summer), emanating from the thick Boston ivy of a university building near my home. Whether in North or in South America, the English house sparrow has found a new world, and one that is its own. Perhaps it is telling us, at the end of the day, that the world is all one, wherever we may be. Wherever we settle for the night, we are all connected, through time and space, to all that the world has been and will be.

A few miles away, darkness settles too on the glacial rock at Salto, Brazil, as on the rocks my canoe has slipped past in the lakes far away in Canada. The Brazilian rock was molded by the forces of nature nearly 300 million years ago, long before the earth had seen the birth of a warm-blooded mammal; the rocks of Canada were shaped by recent glacial ice, in ages that saw people cross from the Old World into the New. The elemental forces of water and ice were the same in the Paleozoic as they are now. Eternal forces of nature have guided the unfolding evolution of our planet, and its plant and animal communities, throughout all time.

For hundreds of millions of orbits around the sun, the earth has intercepted a tiny fraction of the sun's rays as they headed toward outer space, and transmuted their energy into the activities of our living planet. The earth's elements have remained nearly constant, endlessly recycled through the dynamic materials of geologic and biological processes; the fundamental forces working on those materials have not changed from age to age. Yet from such constancy has come not an equilibrium of eternal silence, but a stream of creation that has made our old world forever new.

In a small village in Ohio, the chipmunk inherits a new season from the old and seems to sense the continuity of it all, from its place on the glacial till.

Names of Living
Plants and Animals

The common and scientific names are listed in
the order of their appearance in the text.

introduction

Hemlock	*Tsuga canadensis*
Hay-scented fern	*Dennstaedtia punctilobula*
Cottonwood	*Populus deltoides*
Sycamore (plane tree)	*Platanus occidentalis*
European poplar	*Populus nigra*
Beech (English)	*Fagus sylvatica*
Beech (American)	*Fagus grandifolia*
Yew (English)	*Taxus baccata*

rocks

Veronica	*Veronica officinalis*
Violet	*Viola papilionacea*
Dandelion	*Taraxacum officinale*
English daisy	*Bellis perennis*
Robin (American)	*Turdus migratorius*
Cardinal	*Cardinalis cardinalis*
Loon (great northern diver)	*Gavia immer*
Hummingbird	*Archilochus colubris*

179

water and ice

Red-winged blackbird	*Agelaius phoeniceus*
White pine	*Pinus strobus*
Prickly pear cactus	*Opuntia humifusa*
Sea rocket	*Cakile edentula*
Seaside spurge	*Euphorbia polygonifolia*

people

Maize (corn)	*Zea mays*
Goosefoot (native wild)	*Chenopodium berlandieri*
Goosefoot (garden weed)	*Chenopodium album*
Gourd or squash	*Cucurbita pepo*
Sunflower	*Helianthus annuus*
Marsh elder	*Iva annua*
Beaver	*Castor canadensis*
False hellebore	*Veratrum viride*
Black bear	*Ursus americanus*
Whitetail deer	*Odocoileus virginianus*
Mountain lion (cougar)	*Felis concolor*
Wolf (gray)	*Canis lupus*
Elk	*Cervus canadensis*
Bison	*Bison bison*
Turkey	*Meleagris gallopavo*
Grouse	*Bonasa umbellus*
Quail (bobwhite)	*Colinus virginianus*
Canada goose	*Branta canadensis*
Raccoon	*Procyon lotor*

western reserve

Black ash	*Fraxinus nigra*
Red maple	*Acer rubrum*

Hornbeam	*Carpinus caroliniana*
Elm (American)	*Ulmus americana*
Chestnut (American)	*Castanea dentata*
Tulip tree	*Liriodendron tulipifera*
Black walnut	*Juglans nigra*
White snakeroot	*Eupatorium rugosum*

flora

Ostrich fern	*Matteuccia struthiopteris*
Scouring rush	*Equisetum hiemale*
Polypody	*Polypodium vulgare*
Spruce (red)	*Picea rubens*
Fir (Fraser)	*Abies fraseri*
Woolly adelgid (balsam)	*Adelges piceae*
Red oak	*Quercus rubra*
Hobblebush	*Viburnum alnifolium*
Ponderosa pine	*Pinus ponderosa*
Lodgepole pine	*Pinus contorta*
Englemann spruce	*Picea engelmannii*
Fir (subalpine)	*Abies lasiocarpa*
Yellow birch	*Betula alleghaniensis*
Bur oak	*Quercus macrocarpa*
Sugar maple	*Acer saccharum*
White ash	*Fraxinus americana*
White birch	*Betula papyrifera*
Aspen	*Populus tremuloides*
Bracken	*Pteridium aquilinum*
Phlox	*Phlox* (species)
Sensitive fern	*Onoclea sensibilis*
Cinnamon fern	*Osmunda cinnamomea*
Hickory	*Carya* (species)
Sour gum	*Nyssa* (species)
Dawn redwood	*Metasequoia glyptostroboides*

American plane tree	*Platanus occidentalis*
Asiatic plane tree	*Platanus orientalis*
London plane tree	*Platanus acerifolia*
Oak	*Quercus* (species)
Galax	*Galax aphylla*
Trailing arbutus	*Epigaea repens*
Hepatica	*Hepatica acutiloba*
Trout lily	*Erythronium americanum*
Trillium	*Trillium grandiflorum*
Tamarack	*Larix laricina*
Linden (European)	*Tilia cordata*
Hornbeam (European)	*Carpinus betulus*
Chestnut (European)	*Castanea sativa*
Sycamore maple	*Acer pseudoplatanus*
Norway spruce	*Picea abies*
European larch	*Larix decidua*
Fir (European)	*Abies alba*
Scot's pine	*Pinus sylvestris*
Juniper	*Juniperus communis*
Catalpa	*Catalpa speciosa*
Osage orange	*Maclura pomifera*
Hawthorne	*Crataegus* (species)
Sweetgum	*Liquidambar styraciflua*
Bald cypress	*Taxodium distichum*
Sphagnum moss	*Sphagnum* (species)
Swamp loosestrife	*Decodon verticillatus*
Leatherleaf	*Chamaedaphne calyculata*
Sedge	*Carex* (species)
Cranberry	*Vaccinium macrocarpon*
Pitcher plant	*Sarracenia purpurea*
Sundew	*Drosera rotundifolia*
Buttonbush	*Cephalanthus occidentalis*
Winterberry	*Ilex verticillata*
Marsh fern	*Thelypteris palustris*

Red cedar	*Juniperus virginiana*
Canadian yew	*Taxus canadensis*
Northern white cedar	*Thuja occidentalis*
Basswood	*Tilia americana*
Butternut	*Juglans cinerea*
Beechdrops	*Epifagus virginiana*
Spring beauty	*Claytonia virginica*
Elm (English)	*Ulmus procera*
Bluegrass	*Poa pratensis*
Broad-leaved plantain	*Plantago major*
Nettle	*Urtica dioica*
Dock	*Rumex* (species)
White clover	*Trifolium repens*
Red clover	*Trifolium pratense*
Yarrow	*Achillea millefolium*
Queen Anne's lace	*Daucus carota*
Sheep sorel	*Rumex acetosella*
St. Johnswort	*Hypericum perforatum*
Buttercup	*Ranunculus acris*
Fleabane	*Erigeron* (species)
Strawberry	*Fragaria virginiana*
Beardtongue	*Penstemon* (species)
Goldenrod	*Solidago* (species)
Aster	*Aster* (species)
Winter cress	*Barbarea vulgaris*
Mustard	*Brassica* (species)
Sow thistle	*Sonchus* (species)
Shepherd's purse	*Capsella bursa-pastoris*
Velvet leaf	*Abutilon theophrasti*
Wild buckwheat	*Polygonum convolvulus*
Lambsquarters	*Chenopodium album*
Bindweed	*Convolvulus* (species)
Jimsonweed	*Datura stramonium*
Canada thistle	*Cirsium arvense*

Ragweed	*Ambrosia* (species)
Pokeweed	*Phytolacca americana*
Ironweed	*Vernonia noveboracensis*
Daisy	*Chrysanthemum leucanthemum*
Birdfoot trefoil	*Lotus corniculatus*
Crown vetch	*Coronilla varia*
Sweetclover	*Meliotus* (species)
Teasel	*Dipsacus laciniatus*
Burdock	*Arctium* (species)
Chicory	*Cichorium intybus*
Mullein	*Verbascum thapsis*
Everlasting pea	*Lathyrus latifolius*
Yellow goat's beard	*Tragopogon pratensis*
Nightshade	*Solanum dulcamara*
Rough-fruited cinquefoil	*Potentilla rectum*
Bouncing bet	*Saponaria officinalis*
Speedwell	*Veronica officinalis*
Chickweed	*Stellaria media*
Hop clover	*Trifolium* (species)
Sorrel	*Oxalis corniculata*
Mayweed	*Anthemis cotula*
Self-heal	*Prunella vulgaris*
Moneywort	*Lysimachia nummularia*
Gill-over-the-ground	*Glechoma hederacea*
Dead nettle	*Lamium purpureum*
Deptford pink	*Dianthus armeria*
European bark beetle	*Scolytus multistriatus*
American bark beetle	*Hylurgopinus rufipes*
Gorse	*Ulex europaeus*

fauna

Skunk cabbage	*Symplocarpus foetidus*
Toothwort	*Dentaria diphylla*

Aconite	*Eranthis hyemalis*
Crocus	*Crocus albiflorus*
Scilla	*Scilla* (species)
Daffodil	*Narcissus* (species)
Chipmunk	*Tamias striatus*
Monarch butterfly	*Danaus plexippus*
Shearwater	*Puffinus* (species)
House mouse	*Mus musculus*
Norway rat	*Rattus norvegicus*
White-footed mouse	*Peromyscus leucopus*
Shorttail shrew	*Blarina brevicauda*
Little brown bat	*Myotis lucifugus*
Red squirrel	*Tamiasciurus hudsonicus*
Gray squirrel	*Sciurus carolinensis*
Flying squirrel	*Glaucomys volans*
Elk	*Cervus canadensis*
Moose	*Alces alces*
Red wolf	*Canis rufus*
English house sparrow	*Passer domesticus*
Starling	*Sturnus vulgaris*
Pigeon	*Columba livia*
Ring-necked pheasant	*Phasianus colchicus*
House finch	*Carpodacus mexicanus*
Blue jay	*Cyanocitta cristata*
Chickadee (black-capped)	*Parus atricapillus*
Goldfinch	*Carduelis tristis*
Nuthatch (white-breasted)	*Sitta carolinensis*
Downy woodpecker	*Picoides pubescens*
Carolina wren	*Thryothorus ludovicianus*
Phoebe	*Sayornis phoebe*
Baltimore oriole	*Icterus galbula*
Flicker (yellow-shafted)	*Colaptes auratus*
Junco	*Junco hyemalis*
Tree sparrow (American)	*Spizella arborea*

White-throated sparrow	*Zonotrichia albicollis*
Redpoll	*Carduelis flammea*
Yellow-bellied sapsucker	*Sphyrapicus varius*
Johnny darter	*Etheostoma nigrum*
Banded darter	*Etheostoma zonale*
Variegated darter	*Etheostoma variatus*
Sea lamprey	*Petromyzon marinus*
Alewife	*Alosa pseudoharengus*
Freshwater eel	*Anguilla rostrata*
Goldfish	*Carassius auratus*
Carp	*Cyprinus carpio*
Honeybee	*Apis mellifera*
Housefly	*Musca domestica*
Gypsy moth	*Lymantria dispar*
Catharinea moss	*Catharinea (Atrichum) undulata*
Haircap moss	*Polytrichum commune*
White moss	*Leucobryum glaucum*
Windblown moss	*Dicranum scoparium*
Opossum	*Didelphis marsupialis*
Porcupine	*Erethizon dorsatum*
Armadillo	*Dasypus novemcinctus*
Caribou	*Rangifer caribou*
Bobcat	*Lynx rufus*
Coyote	*Canis latrans*
Wolverine	*Gulo gulo*
Brown bear	*Ursus arctos*
Grizzly bear	*Ursus arctos*
Red fox	*Vulpes vulpes*
Arctic fox	*Alopex lagopus*
Ermine	*Mustela erminea*
Snowshoe hare	*Lepus americanus*
Mountain pika	*Ochotona princeps*
Mountain goat	*Oreamnos americanus*
Bighorn sheep	*Ovidis canadensis*

Pronghorn	*Antilocapra americana*
Muskox	*Ovibos moschatus*
English blackbird	*Turdus merula*
Bobolink	*Dolichonyx oryzivorus*
Red-eyed vireo	*Vireo olivaceus*
Blackburnian warbler	*Dendroica fusca*
Solitary sandpiper	*Tringa solitaria*
Heron (great blue)	*Ardea herodias*
Turkey vulture	*Cathartes aura*
New England aster	*Aster novae-angliae*
Milkweed	*Asclepias* (species)
Poison ivy	*Rhus radicans*
Christmas fern	*Polystichum acrostichoides*
Marginal wood fern	*Dryopteris marginalis*
Waterleaf	*Hydrophyllum appendiculatum*
Wild stonecrop	*Sedum ternatum*
Wood frog	*Rana sylvatica*
Spring peeper	*Hyla crucifer*
Gray treefrog	*Hyla versicolor*
Garter snake	*Thamnophis sirtalis*
Tent caterpillar	*Malacosoma americanum*
Woolly bear	*Pyrrharctia isabella*
Goldenrod gall fly	*Eurosta solidaginis*
Promethea moth	*Callossamia promethea*
Mourning cloak butterfly	*Nymphalis antiopa*
Earthworm	*Lumbricus terrestris*
American toad	*Bufo americanus*
Green frog	*Rana clamitans*
Leopard frog	*Rana pipiens*
Bullfrog	*Rana catesbeiana*
Woodchuck	*Marmota monax*
Red chantrelle	*Cantharellus cinnabarinus*
Amanita	*Amanita* (species)
Willow	*Salix* (species)

Cottontail rabbit	*Sylvilagus floridanus*
Grackle	*Quiscalus quiscula*
Cowbird	*Molothrus ater*
Wood duck	*Aix sponsa*
Bluebird	*Sialia sialis*
Boston ivy	*Parthenocissus tricuspidata*

Bibliography

introduction

Mabey, Richard. *Gilbert White*. London: Century Hutchinson, 1986.

> Everything Mabey writes is interesting, and this is the most detailed biography yet of Gilbert White. Two other short ones are Ronald M. Lockley, *Gilbert White* (London: White Lion, 1976), and Cecil S. Emden, *Gilbert White in His Village* (Oxford: Oxford University Press, 1956). I would recommend reading the first chapter of Mabey's book—or visiting Selborne!—before reading Gilbert White himself.

Thoreau, Henry D. *Walden*. Boston: Ticknor and Fields, 1854.

> In addition to Thoreau's *Walden,* some of the great American nature classics, communicating a sense of place, are Henry Beston, *The Outermost House* (Garden City, N.Y.: Doubleday, Doran, 1928); Aldo Leopold, *A Sand County Almanac* (New York: Oxford University Press, 1949); Joseph Wood Krutch, *The Desert Year* (New York: Sloane, 1952); and Sigurd F. Olson, *The Singing Wilderness* (New York: Knopf, 1956).

White, Gilbert. *The Natural History of Selborne*. Edited by R. M. Lockley. London: Dent, 1949.

> Originally published in 1788, one of the most reprinted books in the English language. My edition is from the Everyman's Library.

rocks

Banks, P. O., and Rodney M. Feldman. *Guide to the Geology of Northeastern Ohio*. Cleveland: Northern Ohio Geological Society, 1970.

Technical descriptions of the Devonian, Mississippian, and Pennsylvanian rocks of Lorain and Ashland counties and eastward, and of glacial till at several locations, as well as a brief history of nineteenth-century geology in northeastern Ohio, are the subjects of this book.

Bownocker, J. A. *Geologic Map of Ohio*. Columbus: State of Ohio Department of Natural Resources, Division of Geological Survey, 1920, 1947, 1981.

About 31 × 38 inches, this color map shows the periods of surface rocks in Ohio.

Case, Gerard R. *A Pictorial Guide to Fossils*. New York: Van Nostrand Reinhold, 1982.

An extensive photographic collection of animal fossils and of some plant fossils.

Continents Adrift. Readings from *Scientific American*, with introductions by J. Tuzo Wilson. San Francisco: Freeman, 1972.

Articles on plate tectonics.

Ernst, J. E., and D. K. Musgrave. "Soil associations of Lorain County." In *An Inventory of Ohio Soils*. Progress Report No. 36. Columbus: State of Ohio Department of Natural Resources, Division of Lands and Soils, 1972.

This article discusses how scientists have classified the dirt beneath Ohio feet.

Hallam, A. *Great Geological Controversies*. Oxford: Oxford University Press, 1983.

Among the controversies interestingly discussed in this book are those concerning continental drift and continental glaciation.

Hubbard, George D. *Dimensions of the Cincinnati Anticline.* Oberlin College Laboratory Bulletin No. 38. Oberlin: Oberlin College, 1893.

A brief historical review and description of the anticline, which was first noted in 1838 by John Locke of the Ohio Geological Survey.

Lafferty, Michael B., ed. *Ohio's Natural Heritage.* Columbus: Ohio Academy of Science, 1979.

A beautifully illustrated introduction to the natural history of Ohio.

La Rocque, Aurele, and Mildred Fisher Marple. *Ohio Fossils.* Bulletin No. 54. Columbus: State of Ohio Department of Natural Resources, Division of Geological Survey, 1955.

Newberry, John S. "Report on the Geology of Lorain County." In *Report of the Geological Survey of Ohio,* edited by John S. Newberry et al., Vol. 2, 206–24. Columbus: Nevin & Myers, 1874.

This early survey shows that the study of rock strata, which began in Britain at the time of the founding of Oberlin, had already been extensively applied to our region by forty years later. By 1874 too, Agassiz's concept of continental glaciation had been used to explain the beach ridges north of Oberlin and the nature of surface deposits on the land.

Petersen, Morris S., J. Keith Rigby, and Lehi F. Hintze. *Historical Geology of North America.* Dubuque, Iowa: Brown, 1973.

There are many excellent textbooks of historical geology; this is a fine short one.

Wegener, Alfred. *The Origin of Continents and Oceans.* 4th ed. Translated by John Biram. 1929. Reprint. New York: Dover, 1962.

Originally published in German in 1915, this book outlined a theory of plate tectonics that was not accepted for more than sixty years.

water and ice

The Beach Ridges of Lorain County. Geology Folder No. 2. LaGrange, Ohio: Lorain County Metropolitan Park District, Board of Park Commissioners.

An excellent map of the beach ridges, from Vermilion and Birmingham on the west, to Avon and North Ridgeville on the east.

Blume, Helmut. *Colour Atlas of the Surface Forms of the Earth.* Translated by Bjorn Wygrala. Edited by Andrew Goudie and Rita Gardner. Cambridge, Mass.: Harvard University Press, 1992.

An extraordinary collection of photographs of landforms, including "roches moutonnees" and other forms resulting from glaciation.

Bretz, J. Harlen. "Correlation of Glacial Lake Stages in the Huron–Erie and Michigan Basins." *Journal of Geology* 72 (1964): 618–27.

Demonstrating that the glacial geology of northern Ohio is not easy or without controversy, this article takes issue with some of Hough's conclusions concerning the postglacial lake stages.

Carney, Frank. "The Abandoned Shorelines of the Oberlin Quadrangle, Ohio" and "The Abandoned Shorelines of the Vermilion Quadrangle, Ohio." *Bulletin of the Scientific Laboratories of Denison University* 16 (1910–1911): 101–17, 233–44.

Detailed descriptions of the beach ridges in the Oberlin region. [Available from the Oberlin College library]

Dobson, Peter. "Remarks on Bowlders." *American Journal of Science* 10 (1826): 217–18.

Dobson argued that rocks on his Connecticut property must once have been suspended in ice and dragged across other rocks.

Flint, Richard Foster. *Glacial and Quaternary Geology*. New York: Wiley, 1971.

A comprehensive (and technical) treatise on the physical and historical geology of glaciers.

Forsyth, Jane L. *The Beach Ridges of Northern Ohio*. Information Circular No. 25. Columbus: State of Ohio Department of Natural Resources, Division of Geological Survey, 1959.

The most recent detailed study of the beach ridges along the entire Ohio shoreline.

Forsyth, Jane L. *Dating Ohio's Glaciers*. Information Circular No. 30. Columbus: State of Ohio Department of Natural Resources, Division of Geological Survey, 1961.

A report of results of carbon-14 analyses done on wood and other materials from glacial deposits in Ohio.

Granger, Ebenezer. "Notice of a Curious Fluted Rock at Sandusky Bay." *American Journal of Science* 6 (1823): 180.

Ten years before the founding of Oberlin, and thirteen years before Louis Agassiz began speculating about glaciers, Granger argued that the grooves on this Ohio rock must have been produced by abrasion from a hard body.

Henderson, Lawrence J. *The Fitness of the Environment*. New York: Macmillan, 1913.

A classic statement of how the fundamental characteristics of simple substances, especially water, are related to the requirements of life as we know it.

Hough, Jack L. *Geology of the Great Lakes*. Urbana: University of Illinois Press, 1958.

A brief description of the rocks and topography of the Great Lakes, and a detailed account of the glacial and postglacial history of the lakes.

193

Kurten, Bjorn. *The Ice Age*. New York: Putnam, 1972.

A nicely illustrated and very readable account by a distinguished Swedish paleontologist.

Leverett, Frank. *Glacial Formations and Drainage Features of the Erie and Ohio Basins*. Monograph No. 41. United States Geological Survey. Washington, D.C.: Government Printing Office, 1902.

This massive survey covers Ohio and parts of surrounding states, including western New York, with many maps.

Leverett, Frank, and F. B. Taylor. *The Pleistocene of Indiana and Michigan and the History of the Great Lakes*. United States Geological Survey Monograph No. 53. Washington, D.C.: Government Printing Office, 1915.

Extending Leverett's previous study to the west, Leverett and Taylor attempted to correlate the postglacial stages of the Erie, Huron, and Michigan basins. Their survey has served as a foundation for all subsequent studies.

Martin, Helen M. *"Ne-Saw-Je-Won," as the Ottawas Say: A Tale of the Waters that Run Down from Lake Superior to the Sea*. Cleveland: Feather, 1939.

A more popular account of the history of the Great Lakes.

Pielou, E. C. *After the Ice Age: The Return of Life to Glaciated North America*. Chicago: University of Chicago Press, 1991.

This masterful summary, engagingly written and attractively illustrated, of recent studies of the postglacial history of North America also contains an extensive bibliography of the relevant scientific literature.

Sharp, Robert P. *Living Ice: Understanding Glaciers and Glaciation*. Cambridge: Cambridge University Press, 1988.

This introduction to the physical geology of glaciers, with magnificent photographs, discusses the "roches moutonnees" (whaleback rocks, rock drumlins) produced by glaciers.

Stuckey, Ronald L. "William George Tight and the Naming of the Teays River and Lake Tight." In *Teays-Age Drainage Effects on Present Distributional Patterns of Ohio Biota*, edited by Charles C. King, 7-14. Ohio Biological Survey Information Circular No. 11. Columbus: Ohio State University, 1983.

An account of the history of the discovery of the preglacial Teays River.

White, George W. *Glacial Geology of Northeastern Ohio*. Bulletin No. 68. Columbus: State of Ohio Department of Natural Resources, Division of Geological Survey, 1982.

A beautiful 32- × 34-inch color map, together with a technical description of glacial features in Lorain, Ashland, and Richland counties, and eastward (does not include Erie and Huron counties).

people

Bakeless, John. *The Eyes of Discovery: The Pageant of North America as Seen by the First Explorers*. Philadelphia: Lippincott, 1950.

What the North American land looked like to the first Europeans.

Baldwin, C. C. "The Iroquois in Ohio" (Tract 40) and "Early Indian Migration in Ohio" (Tract 47). In *Western Reserve Historical Society*. vol. 2, *Tracts* 37–72. Cleveland: 1888.

Describes how the Iroquois established control over northern Ohio from the middle of the seventeenth century to the end of the eighteenth century, and speculates on migrations into Ohio during that time.

Carrier, Lyman. *The Beginnings of Agriculture in America*. New York: McGraw-Hill, 1923.

An early work recognizing, and describing in interesting detail, the extensive agriculture carried out by the Native American people.

Claiborne, Robert. *The First Americans*. New York: Time-Life Books, 1973.

A beautifully illustrated introduction to the history of the Native people of North America.

Converse, Robert N. *Ohio Flint Types*. Rev. ed. Columbus: Archaeology Society of Ohio, 1973.

Drawings and descriptions of flint implements that have been found in or near Ohio.

Cronon, William. *Changes in the Land: Indians, Colonists, and the Ecology of New England*. New York: Hill and Wang, 1983.

A brilliant study of what the New England land was like when the Native Americans lived there, and how it changed with colonization. Much of the material is relevant to an interpretation of conditions in Ohio.

Crosby, Alfred W. *The Columbian Exchange: Biological and Cultural Consequences of 1492*. Westport, Conn.: Greenwood Press, 1972.

Crosby, Alfred W. *Ecological Imperialism. The Biological Expansion of Europe, 900–1900*. Cambridge: Cambridge University Press, 1986.

History from an ecological point of view: Crosby has portrayed in rich detail the ecological effects of the human contact between Europe and America.

Denevan, William M. "The Pristine Myth: The Landscape of the Americas in 1492." *Annals of the Association of American Geographers*. 82 (1992): 369–85.

The most recent scholarship, confirming that where Native Americans farmed (as they did in northern Ohio until the seventeenth century), the land was not a wilderness.

Densmore, Frances. *How Indians Use Wild Plants for Food, Medicine and Crafts.* New York: Dover, 1974. [Reprint of *Uses of Plants by the Chippewa Indians.* Washington, D.C.: Smithsonian Institution, 1926–1927]

The pioneering study by a graduate of the Oberlin Conservatory of Music.

Early Man in America. Readings from *Scientific American,* with an introduction by Richard S. MacNeish. San Francisco: Freeman, 1973.

Articles on the archeology of the Paleo-Americans.

Farb, Peter. *Man's Rise to Civilization as Shown by the Indians of North America from Primeval Times to the Coming of the Industrial State.* New York: Dutton, 1968.

An excellent survey of Native American history.

Jennings, Jesse D., ed. *Ancient North Americans.* New York: Freeman, 1983.

A fine survey of archeological findings from representative sites throughout North America.

MacGowan, Kenneth, and Joseph A. Hester, Jr. *Early Man in the New World.* New York: American Museum of Natural History and Doubleday, 1962.

An engaging discussion of American history, from the Paleo-Americans to the invention of agriculture.

McKenzie, Douglas H. and John E. Blank. "The Eiden Site: Late Woodland from the South-Central Lake Erie region." In *The Late Prehistory of the Lake Erie Drainage Basin,* edited by David S. Brose, 305–26. Cleveland: Cleveland Museum of Natural History, 1976.

A summary of studies done at this site at the junction of French Creek and the Black River.

McNeill, William H. *Plagues and Peoples*. Garden City, N.Y.: Doubleday, Anchor, 1976.

A lucid exposition of the effects of disease in history, especially the effects of the European diseases on the Natives of the New World.

Mills, William C. "Excavations of the Adena Mound." In *Ohio Archeological and Historical Publications*, Vol. 10, 451–79. Columbus: Ohio State Archeological and Historical Society, 1902.

The original report of the excavations of the site that gave the Adena period its name.

Parkman, Francis. *LaSalle and the Discovery of the West*. 1869. Boston: Little, Brown, 1907.

A vivid portrayal of La Salle's explorations from the St. Lawrence, through the southern Great Lakes, and into the Mississippi.

Quimby, George Irving. *Indian Life in the Upper Great Lakes, 11,000 B.C. to A.D. 1800*. Chicago: University of Chicago Press, 1960.

Describes the relations between a changing postglacial environment and the ecology of Native peoples in the region to the north and west of Ohio.

Russell, Howard S. *Indian New England before the Mayflower*. Hanover, N. H.: University Press of New England, 1980.

An admirable and interesting study of American life in the years before 1620. While the area of study is New England, much of it can be extrapolated to life in the lower Great Lakes region.

Sale, Kirkpatrick. *The Conquest of Paradise: Christopher Columbus and the Columbian Legacy*. New York: Plume, 1990.

A brilliant presentation based on extensive scholarship and critical reevaluation of sources.

Sauer, Carl Ortwin. *Sixteenth-Century North America: The Land and the Peo-*

ple as Seen by the Europeans. Berkeley: University of California Press, 1971.

A discussion of the early European view of North America.

Shane, Orrin C., III. "The Leimbach Site: An Early Woodland Village in Lorain County, Ohio." In *Studies in Ohio Archeology*, edited by Olaf H. Prufer and Douglas H. McKenzie, 98–120. Cleveland: Press of Western Reserve University, 1967.

Report of the archeological studies done at this important site on the Vermilion River.

Smith, Bruce D. *The Emergence of Agriculture*. New York: Scientific American Library, 1995.

Smith presents recent evidence that eastern North America was a center for the development of agriculture: that important food plants were domesticated there more than 2,000 years before maize, originating in Mexico, became adapted to its climates.

Snow, Dean. *The Archeology of North America*. New York: Viking, 1976.

Excellent photographs of artifacts of Native Americans.

Tanner, Helen Hornbeck. *Atlas of Great Lakes Indian History*. Norman: University of Oklahoma Press, 1987.

Attractive new maps and interesting summaries of what is known of the history of the Native Americans who inhabited our Great Lakes region.

Vietzen, Raymond C. *The Immortal Eries*. Elyria, Ohio: Wilmot, 1945.

Vietzen grew up in Lorain County, and has spent a long lifetime in archeological studies of the Native people. This early work, based on extensive fieldwork by the author, describes the major archeological sites (including the Leimbach and Franks sites on the Vermilion River) known in northern Ohio in 1945, and presents a thoughtful and empathetic view of the people who once inhabited

199

the southern shores of Lake Erie. It contains many photographs of artifacts and skeletons.

Whittlesey, Charles. *Early History of Cleveland.* Cleveland: Fairbanks, Benedict, 1867.

The Seneca story of how the Iroquois people destroyed the Eries in 1655 is recounted here.

Wilcox, Frank N. *Ohio Indian Trails.* Kent, Ohio: Kent State University Press, 1970.

A map and descriptions of major trails used by the native Ohioans.

Williams, Michael. *Americans and Their Forests: A Historical Geography.* Cambridge: Cambridge University Press, 1989.

The management of forests by the Native Americans is discussed.

western reserve

Boynton, Washington Wallace. *The Early History of Lorain County.* Cleveland: Western Reserve Historical Society, 1892.

Further notes on the surveying and division of Lorain County.

Ellis, William Donohue. *The Cuyahoga.* New York: Holt, Rinehart and Winston, 1966. Reprint. Dayton: Landfall Press, 1975.

For a little more about the trials of the surveyors, see Chapter 5.

Fairchild, James Harris. *Early Settlement and History of Brownhelm.* Oberlin, 1867. Reprint. *Country Boy: Growing Up in Northern Ohio in the 1820's.* Edited by Geoffrey Blodgett. Elyria, Ohio: Lorain County Historical Society, 1993.

Written for the fiftieth anniversary of the founding of Brownhelm, this is surely one of the most interesting accounts anywhere of life and land in the southern Great Lakes in the early nineteenth century.

Fairchild, James Harris. *Oberlin: The Colony and the College*. Oberlin: Goodrich, 1883.

> James Fairchild and his brother, Henry, became students at Oberlin in 1834, and James became president of the College in 1866 (Henry became president of Berea College). His history and descriptions of the land are firsthand.

Fletcher, Robert Samuel. *A History of Oberlin College from Its Foundation Through the Civil War*. Oberlin: Oberlin College, 1943.

> Immensely detailed and beautifully presented history of the founding of Oberlin. Fletcher's wide interests did not, however, extend very far into environmental matters; his range and township numbers for Russia Township (containing Oberlin) are off by one.

Foskett, Helen Robinette. *History of New London, Ohio*. New London, Ohio: New London Public Library, 1976.

> Interesting notes on the founding and early history of this community in the Firelands, southwest of Oberlin.

Hatcher, Harlan. *The Western Reserve: The Story of New Connecticut in Ohio*. Rev. ed. Cleveland: World, 1966.

> Chapters 1–8 give an interesting summary of the surveying and settlement of the Western Reserve.

Hedrick, Ulysses Prentiss. *A History of Agriculture in the State of New York*. Albany, N.Y.: Lyon, 1933.

> Chapter 11 describes how the Erie Canal facilitated the transport of agricultural goods across New York State, aiding the southern Great Lakes, but causing the abandonment of many New England farms.

Horton, John J. *The Jonathan Hale Farm: A Chronicle of the Cuyahoga Valley*. The Western Reserve Historical Society Publication No. 116. Cleveland: Western Reserve Historical Society, 1961.

A glimpse of very early farm life on the Western Reserve. The farm is today a museum.

Howse, Derek. *Greenwich Time and the Discovery of Longitude.* Oxford: Oxford University Press, 1980.

A history of the observatory that gave the world its longitudes.

Phillips, Wilbur H. *Oberlin Colony, the Story of a Century.* Oberlin, 1933.

Contains interesting remarks on the land of Lorain County at the time of the founding of the college.

Swing, Albert Temple. *James Harris Fairchild.* New York: Revell, 1907.

Chapters 1–3 of this biography describe the Fairchild family's move from Stockbridge, Massachusetts, to the Western Reserve.

Thomson, Betty Flanders. *The Changing Face of New England.* Boston: Houghton Mifflin, 1977.

A delightful history of the land in New England. Of special relevance is her description of the difficulties of New England farms in the early nineteenth century.

Whittlesey, Charles. *Fugitive Essays upon Interesting and Useful Subjects Relating to the Early History of Ohio.* Hudson, Ohio: Sawyer, Ingersoll, 1852.

Whittlesey was one of the original settlers of Brownhelm, and his description of the beauty of the beach ridges is interesting. He is not the Charles Whittlesey, Ohio geologist, for whom the postglacial Lake Whittlesey is named.

Williams, W. W. *History of Lorain County, Ohio.* Philadelphia: Williams, 1879.

Gives much attention to the surveying and apportionment of the land (and to local geology and paleontology).

flora

Braun, E. Lucy. *Deciduous Forests of Eastern North America.* Philadelphia: Blakiston, 1950. Reprint. New York: Hafner, 1967.

From extensive field studies, it classifies and characterizes the botanical regions of the eastern forest. Northern Ohio is classified as part of the beech–maple forest region. A large map is included.

Braun, E. Lucy. "A History of Ohio's Vegetation." *Ohio Journal of Science* 34 (1934): 247–57.

From early studies done in Ohio and elsewhere, by herself and by Paul Sears at Oberlin College, Braun sketches the probable postglacial botanical history of Ohio.

Claypole, E. W. "Traces of the Ice Age in the Flora of the Cuyahoga Valley" (Tract 84) In *Western Reserve Historical Society.* Vol. 3, Tracts 73–84. Cleveland: Western Reserve Historical Society, 1892.

An interesting early recognition that glacial history had left its mark on the distribution of plants in northern Ohio.

Dupree, A. Hunter. *Asa Gray.* Cambridge, Mass: Harvard University Press, 1959.

This biography of the pioneering American botanist describes his dispute with Agassiz concerning the interpretation of the distribution of plants in America and Eurasia.

Farb, Peter. *Face of North America: The Natural History of a Continent.* New York: Harper & Row, 1963.

A superb introduction to the geology and ecology of North America. If one could read only one book about the land of North America, this might be the one to pick.

Gleason, Henry A., and Arthur Cronquist. *The Natural Geography of Plants.* New York: Columbia University Press, 1964.

A detailed but nontechnical exposition of the factors affecting the geographic distribution of plants.

Gordon, Robert B. *Natural Vegetation of Ohio at the Time of the Earliest Land Surveys.* Ohio Biological Survey. Columbus: Ohio State University, 1966.

A beautiful 31- × 33-inch color map, with sources of the data and brief descriptions of the vegetation types.

Gordon, Robert B. *The Natural Vegetation of Ohio in Pioneer Days.* Bulletin of the Ohio Biological Survey, n.s., vol. 3, no. 2. Columbus: Ohio State University, 1969.

A magnificent attempt to reconstruct what the flora of the land was like when settlers first came into Ohio from Virginia and New England. It has long been assumed that the vegetation as it was in 1800 was the "natural" vegetation of the land. It is likely, however, that the vegetation of 1800 was considerably different from that of 1500, when normal populations of the Native Americans (undiminished yet by European diseases) were farming the land.

Graves, Arthur Harmount. *Illustrated Guide to Trees and Shrubs.* Rev. Ed. New York: Harper, 1956. Reprint. New York: Dover, 1992.

Outstanding illustrations of winter twigs as well as leaves, for learning the trees at any season. Long out of print, it is back in print at a bargain price.

Hylander, Clarence J. *Wildlife Communities: From the Tundra to the Tropics in North America.* Boston: Houghton Mifflin, 1966.

A nontechnical presentation of the plants and animals of the major biomes of North America.

Jones, George T. *Tappan Square.* Oberlin: Oberlin College, 1987.

A map of the individual trees of Tappan Square, giving both common and scientific names.

Lindeman, Karl. *Forest Resources of Lorain County, Ohio.* Forestry Publication No. 68. Wooster: Ohio Agricultural Experiment Station, Division of Forestry, 1940.

A survey of the woodlands of Lorain County as they were in 1940.

Mossman, Ronald Eugene. "A Floristic and Ecological Evaluation of Camden (Bog) Lake, Lorain County, Ohio." Master's thesis, Ohio State University, 1972.

The recovery of the bog, following its earlier drainage by the city of Oberlin, can be evaluated by comparisons with this valuable study. [Copy available in the Oberlin College library]

Peattie, Donald Culross. *Flowering Earth.* New York: Putnam, 1939.

Poetry and science combined. Surely one of the most beautiful books ever written about the flora of the land.

Peterson, Roger Tory, and Margaret McKenny. *A Field Guide to Wildflowers of Northeastern and North-Central North America.* Boston: Houghton Mifflin, 1968.

With superb drawings by the authors, this guide covers a range with northern Ohio in the center. My first copy is nearly worn out.

Platt, Rutherford. *The Great American Forest.* Englewood Cliffs, N.J.: Prentice-Hall, 1965.

A fine popular account of the history and nature of American forests.

Potter, Loren D. "Post-Glacial Forest Sequence of North Central Ohio." *Ecology* 28 (1947): 396–417.

A study of pollen from fifteen bogs of northern Ohio, showing sequences of trees similar to those found by Sears.

Ralston, Chester F. *Oberlin Trees of Campus and Town.* Oberlin: Oberlin College, 1944.

A fascinating history of Oberlin's trees, from 1833 to 1944, with a beautiful foreword by Donald M. Love: "It would be hard to find a more satisfactory symbol for the life and growth of a college or community than a tree—deep-rooted and far-branching, reaching down into the soil of the past for strength and nourishment, reaching up into the clean, bright air of the future in aspiration and hope, reaching out in friendly shade and protection over the vigorous life of the present."

Sears, Paul B. "The Natural Vegetation of Ohio, I: A Map of the Virgin Forest." *Ohio Journal of Science* 25 (1925–1926): 139–49.

Sears, Paul B. "The Natural Vegetation of Ohio, II: The Prairies." *Ohio Journal of Science* 26 (1925-1926): 128–46.

Sears, Paul B. "The Natural Vegetation of Ohio, III: Plant Succession." *Ohio Journal of Science* 26 (1925-1926): 213–31.

From the records of the original surveyors of Ohio, Sears attempted to map the types of forests, and the locations of prairies, in the Ohio of 1800.

Sears, Paul B. "Pollen Analysis of Mud Lake Bog in Northern Ohio." *Ecology* 12 (1931): 605–55.

Sears, Paul B. "A Record of Post-Glacial Climate in Northern Ohio." *Ohio Journal of Science* 30 (1930): 205–17.

The first pollen analyses made from two bogs of northern Ohio (Bucyrus Bog to the west of Oberlin, and Mud Lake Bog to the southeast) showed a postglacial sequence from fir and spruce to pine and to deciduous trees, indicating a gradual warming of climate and reestablishment of trees from farther south.

Secrest, Edmund, and J. S. Houser. "Report of a Study of Shade Trees and Shade Tree Conditions, and Recommendations for a Policy of Future Management of the Trees of the City of Oberlin."

A study done in the early 1940s by a forester and an entomologist from the Ohio Agricultural Experiment Station in Wooster. [Typescript available in the Oberlin College library]

Shelford, Victor E. *The Ecology of North America*. Urbana: University of Illinois Press, 1963.

A detailed treatise on the biomes of North America, by the man who studied the Lake Michigan dunes with Henry Chandler Cowles.

Strobel, Gary A., and Gerald N. Lanier. "Dutch Elm Disease." *Scientific American*, August 1981, 56–66.

An interesting account of the nature and history of this disease, which has so greatly affected the trees of Oberlin (and those of eastern North America and Europe).

Stuckey, Ronald L. "Origin and Development of the Concept of the Prairie Peninsula." In *The Prairie Peninsula—In the "Shadow" of Transeau*, edited by Ronald L. Stuckey and Karen J. Reese, 4–23. Ohio Biological Survey, Biological Notes No. 15. Columbus: Ohio State University, 1981.

An interesting history of the idea that prairies extended into Ohio in postglacial times.

Thoreau, Henry D. *Faith in a Seed*. Edited by Bradley P. Dean. Washington, D.C.: Island Press, 1993.

The publication of "The Dispersion of Seeds" and other late natural history writings of Thoreau, showing the wealth of his ecological observations and insights.

Watts, May Theilgaard. *Reading the Landscape of America*. [revised and expanded edition of *Reading the Landscape* (1955)]. New York: Macmillan, 1975.

Chatty but highly informed: how to interpret what one sees in the flora of different places. For visitors to Europe, her *Reading the Landscape of Europe* (New York: Harper & Row, 1971) is even better.

fauna

Allen, Durward L. *Our Wildlife Legacy.* Rev. ed. New York: Funk and Wagnalls, 1974.

Essays on the interaction of humanity with wildlife in North America.

Baker, Frank Collins. *The Life of the Pleistocene or Glacial Period.* Urbana: University of Illinois Press, 1920.

Describes a number of the early fossil finds in Ohio and elsewhere.

Black River Audubon Society. *Wing Tips.* Elyria, Ohio.

Listings of bird sightings, published periodically.

Burns, Noel M. *Erie: The Lake That Survived.* Totowa, N.J.: Rowman & Allanheld, 1985.

A description of the physiography and biology of Lake Erie, and of the pollution problems it has suffered.

Burt, William Henry. *Mammals of the Great Lakes Region.* Ann Arbor: University of Michigan Press, 1957.

Technical descriptions, and excellent range maps.

Cahalane, Victor H. *Mammals of North America.* New York: Macmillan, 1961.

Rich in detail and nontechnical, with magnificent drawings by Francis Lee Jaques.

Colbert, Edwin H., ed. cons. *Our Continent: A Natural History of North America.* Washington, D.C.: National Geographic Society, 1976.

A beautiful volume describing the biological and geologic history of North America.

Cowles, Henry Chandler. "The Ecological Relations of the Vegetation on the

Sand Dunes of Lake Michigan." *Botanical Gazette* 27 (1899): 95–391. Reprinted in part in *Foundations of Ecology: Classic Papers with Commentaries*, edited by Leslie A. Real and James H. Brown. Chicago: University of Chicago Press, 1991.

A pioneering study of ecological succession on the Lake Michigan dunes.

Dawson, W. L. *The Birds of Ohio*. 2 cols. Columbus: Wheaton, 1903.

The pioneering work.

Devoto, Bernard. *The Course of Empire*. Boston: Houghton Mifflin, 1952.

A vivid summary of the exploration of North America, the fur trade, and the waterways used for exploration and trade.

Feare, Christopher. *The Starling*. Oxford: Oxford University Press, 1984.

An interesting summary of the biology of the European bird that has become the most numerous bird in Ohio.

Hay, Oliver P. *The Pleistocene of North America and Its Vertebrated Animals from the States East of the Mississippi River and from the Canadian Provinces East of Longitude 95 Degrees*. Washington, D.C.: Carnegie Institution, 1923.

An invaluable listing, with maps, of pre-1923 fossil finds in the eastern states, including Ohio.

Hulbert, Archer Butler. *Historic Highways of America*. Vol. 1, *Paths of the Mound-Building Indians and Great Game Animals*. Cleveland: Clark, 1902.

An account of how animal trails became paths of the Native people, and then roads of later settlers.

Johnson, Perry F. *Bird Check List*. LaGrange, Ohio: Board of Park Commissioners.

A pamphlet listing birds that have been seen in Lorain County, Ohio. Other pamphlets are available from the Lorain County Metropolitan Park District for the turtles, snakes, frogs and toads, and common fishes of Lorain County.

Jones, Lynds. *Spring Bird Migrations at Oberlin, Ohio 1896–1935.* Oberlin: privately printed, 1935.

An earlier airline schedule for Oberlin. [Available in the Oberlin College library]

Kurten, Bjorn. *Before the Indians.* New York: Columbia University Press, 1988.

A popular account of mammalian evolution in North America in the late Tertiary and Quaternary, with beautiful drawings.

Kurten, Bjorn, and E. Anderson. *Pleistocene Mammals of North America.* New York: Columbia University Press, 1980.

A technical, encyclopedic survey of 562 fossil mammalian species from Pleistocene times in North America.

Marchand, Peter J. *Life in the Cold: An Introduction to Winter Ecology.* 2nd ed. Hanover, N.H.: University Press of New England, 1991.

A brilliant summary of the winter adaptations of plants and animals.

Matthiessen, Peter. *Wildlife in America.* New York: Viking, 1959.

The recent history of American wildlife, vividly portrayed by a naturalist–novelist.

Morse, Eric W. *Fur Trade Canoe Routes of Canada/Then and Now.* 2nd ed. Toronto: University of Toronto Press, 1979.

Morse has retraced in his canoe the major routes used between Hudson Bay or the St. Lawrence River and Lake Athabasca in the Northwest. He writes with brilliant clarity about the economy of the fur trade.

Mumford, Russell E., and John O. Whitaker. *Mammals of Indiana*. Bloomington: Indiana University Press, 1982.

Detailed, technical descriptions, and maps of sightings in the counties of Indiana.

Newman, Donald L. *A Field Book of Birds of the Cleveland Region*. Cleveland: Cleveland Museum of Natural History, 1969.

Like an airline schedule in graphic form, depicts times of arrival and departure, and periods of residence.

Ordish, George. *The Year of the Butterfly*. New York: Scribner, 1975.

A charming account of the natural history of the monarch, presented as much as possible from the butterfly's point of view.

Orr, Robert T. *Mammals of North America*. New York: Doubleday, n. d.

A brief survey of present North American mammals, with photographs.

Peterjohn, Bruce G. *The Birds of Ohio*. Bloomington: Indiana University Press, 1989.

With beautiful paintings by William Zimmerman, this is the only comprehensive book of Ohio birds since Dawson's book of 1903.

Ray, Arthur J. *Indians in the Fur Trade: Their Role as Hunters, Trappers and Middlemen in the Lands Southwest of Hudson Bay, 1660–1870*. Toronto: University of Toronto Press, 1974.

A detailed and fascinating discussion, much of which is relevant to the lower Great Lakes area.

Rue, Leonard Lee, III. *The World of the Beaver*. Philadelphia: Lippincott, 1964.

A readable account of the biology of the beaver, nicely illustrated with photographs. Other titles in this series describe the opossum, the whitetail deer, and many other northern Ohio vertebrates.

Schorger, A. W. *The Passenger Pigeon: Its Natural History and Extinction.* Madison: University of Wisconsin Press, 1955. Reprint. Norman: University of Oklahoma Press, 1973.

A sad story, fully and vividly told.

Shelford, Victor E. *Animal Communities in Temperate America.* Chicago: University of Chicago Press, 1913.

Includes a description of the author's study of the Lake Michigan dunes.

Simpson, George Gaylord. *The Geography of Evolution.* Philadelphia: Chilton Books, 1965.

Essays by one who contributed greatly to the modern understanding of the evolution and biogeography of vertebrates.

Stanley, Steven M. *Extinction.* New York: Freeman, 1987.

A discussion of the great extinctions seen in the fossil record and their possible causes.

Summers-Smith, D. *The House Sparrow.* London: Collins, 1963.

The biology of one of the modern world's most successful birds: the "English" sparrow, which is now also the American (and everybody else's) sparrow.

Trautman, Milton B. *The Fishes of Ohio.* Rev. ed. Columbus: Ohio State University Press, 1981.

A work of extraordinary scholarship and beauty, with detailed maps of the distribution of fish species in Ohio.

Trautman, Milton B. *The Ohio Country from 1750 to 1977—A Naturalist's View.* Ohio Biological Survey, Biological Notes No. 10. Columbus: Ohio State University, 1977.

An incisive general summary of how the Ohio land has changed, with observations too on individual animal species.

Tyning, Thomas F. *A Guide to Amphibians and Reptiles*. Boston: Little, Brown, 1990.

An excellent recent summary of the natural history of representative North American amphibians and reptiles, including a discussion of their winter adaptations.

Urquhart, Fred A. *The Monarch Butterfly*. Toronto: University of Toronto Press, 1960.

The biology of the monarch butterfly as known in 1960.

Urquhart, Fred A. *The Monarch Butterfly: International Traveler*. Chicago: Nelson Hall, 1987.

A fascinating account of Urquhart's fifty years of study of the monarch's biology and migrations, including the discovery of the wintering site in Mexico.

Williams, Arthur B., ed. *Birds of the Cleveland Region*. Cleveland: Cleveland Museum of Natural History, 1950.

A listing of all species seen in Cleveland from 1800 to 1950—within a 30-mile radius of Public Square.